PRACTICAL
HANDGUN BALLISTICS

PRACTICAL HANDGUN BALLISTICS

By

MASON WILLIAMS

Firearms and Ballistic Consultant
Police Firearms Training Instructor
Life Member, National Rifle Association
Life Member, United States Revolver Association
Outdoor Writer's Association of America

CHARLES C THOMAS • PUBLISHER

Springfield • Illinois • U.S.A.

Published and Distributed Throughout the World by

CHARLES C THOMAS • PUBLISHER
Bannerstone House
301-327 East Lawrence Avenue, Springfield, Illinois, U.S.A.

ISBN 0-398-04032-X

Library of Congress Catalog Card Number: 79-24667

With THOMAS BOOKS *careful attention is given to all details of manufacturing and design. It is the Publisher's desire to present books that are satisfactory as to their physical qualities and artistic possibilities and appropriate for their particular use.* THOMAS BOOKS *will be true to those laws of quality that assure a good name and good will.*

Printed in the United States of America
V-00-2

Library of Congress Cataloging in Publication Data

Williams, Mason.
Practical handgun ballistics.

1. Ballistics. 2. Pistols. 3. Cartridges.
I. Title.
UF820.W54 623.4′426 79-24667
ISBN 0-398-04032-X

INTRODUCTION

MANY thousands of years ago somewhere on earth, a man stalked a succulent lizard sunning himself on a rock. The man held a small, rounded, worn stone in his right hand. Abruptly, he moved and brought the stone down onto the top of the lizard's head, crushing it.

Somewhere else, another man had lain in wait for hours. Now, the single line of deerlike creatures moved below the rock ledge. With sweat pouring down his body from the strain, the man slowly raised the large rock above his head and flung it down upon the deer creature, breaking its back. Other men leapt out from hiding places, their sharpened sticks readied for the death blow.

Somewhere else, another man studied the rocks at his feet and then bent down to pick up a smooth, round stone that fit his hand. His large eyes moved to the small, four-legged creature that crouched, ready to move at the slightest motion, about fifteen feet away. Slowly, the man drew back his arm and then, with tremendous force, flung the stone at the animal, killing it instantly.

Note that each incident required an evaluation of the situation, — ballistics. Throughout the ages, ballistics have played a very important part in the evolution of civilization. No matter how terrible the weapon of war, it still had to be moved in some manner from the sender to the target, whether it was the horrible Greek fire, the Khan's war elephants, or projectiles in all their varied shapes and types. Remember that spears, arrows, stones, clubs had to be delivered.

In each and every case, someone had to figure out just how the projectile could be delivered to the target. In earliest days, this proved relatively simple. Then came gunpowder and the study of ballistics as we know it today. Firearms in all their phases changed warfare and created problems that we continue to try to solve. The man knew he could fling the stone or hurl the rock only so far. Today, we demand results from ammunition and handguns that are unrealistic.

The study of ballistics is the study of the projectile in three places: within the handgun, in the air, and from the moment it strikes the target to the time it comes to rest. We cannot intelligently understand, or attempt to evaluate, a projectile until we have carefully examined to what the projectile has been subjected. It is one thing to pick up a spent bullet, look at it, state that it is a .38 Special, 158 grain, jacketed, probably hollow-point, soft-nose bullet and that it expanded to about caliber .65. This is fine, but a ballistician might scrutinize the same bullet and comment that if it had been fired from a revolver at the estimated distance, it could never have expanded to .65 caliber or be in the condition that it now is. In other words, something is not quite right. This is one of the problems encountered in ballistics: there are so few constants that it is nearly impossible to make many specific statements. Conditions that are right at eleven feet are all wrong at sixty feet. Distances vary, barrel lengths may run from one and one-half inches up to eight or nine inches. Tolerances within identical handguns can affect bullet performance, to a point. The condition (both mental and physical) of the target often controls whether it drops or continues to move. A load and a handgun that perform flawlessly month after month at fifteen yards may fail completely the next day due to basic alterations in the composition and make-up of the target, all other details remaining the same.

This is why handgun and ammunition manufacturers are continually carrying out evaluations and experiments in an attempt to "discover" the ideal load for law enforcement and for hunting. Actually the hunter has the better deal, because his targets usually fall within prescribed sizes and weights. He can stalk, choose his time and place, and fire when prepared. The police officer must take what he finds and do the best with what he has under conditions usually controlled by the criminal.

While it is true that ballistics is the study of the projectile, there are many other factors that must also be considered. If we could place the cartridge by itself, isolated from people and mechanical things, ballistics would be a true study of the projectile. Unfortunately, we cannot toss the cartridge into the air, cry "fire," and have it ignite and propel the bullet to the target. The cartridge must be enclosed so the propellants and pressures can be retained to exert all their forces upon the base of the bullet. The bullet must then be so located that it will be guided at an ever-increasing velocity in the direction of the target. Because of this, we must evaluate and examine everything that goes on within a handgun, not only to ignite the primer but also to start the powder burning. There are many details within the handgun that can affect the bullet, the cartridge case, and the primer.

Similarly, when the bullet leaves the muzzle of the handgun, it enters air to become influenced by an entirely new set of factors, such as wind, distance, and obstructions. When the bullet finally arrives at the target, it must be capable of dealing with the target composition. Many bullets are ideal for punching holes in paper, while others have been constructed to penetrate iron plate. What if the present target is neither paper nor iron plate?

Because of all this, we have interior ballistics, exterior ballistics, and terminal ballistics. Not until we understand these basics can we pick up a bullet, examine it, and make some good guesses. Only by examining thousands of bullets of different design, weights, and construction that have been fired from varying distances into greatly differing targets can we make an educated guess. Not until then can we say this bullet base should be rounded or concave, this bullet nose opened up slightly, the jacket fractured, or the lead composition was made harder with antimony. There are no pat answers in this ballistic business.

I hope this book will open up to many people the fascinating challenges that exist in ballistics. The unimaginative will be bored with the lack of specific answers and explanations, but those with searching, open, questioning minds will discover an entirely new world. It is time that the dull ballistic tables are put into their rightful place, just as a rock fitted into a fireplace. The rock is only a dull item within a huge structure that will, when completed, provide warmth, color, relaxation, and pleasure. The same is true of ballistics when knowledge is based upon tests, evaluations, shooting, handloading, and the whole range of empirical handgun experience. The knowledgeable could then say "this load will not put this bullet through the target at forty feet, but if we increase the powder charge, decrease bullet weight, and move up four feet, this load could be precisely what we are looking for." Being able to cut through hundreds of pages of tables and charts, put together by men with half a dozen degrees but who have never shot a firearm in their lives, would provide law enforcement with long-sought information. Being able to glance at a page of charts and instantly pick out the key problem and explain why — even though these charts are correct — they do not apply to the situation at hand can save money and time.

I am directing this book at the average person in the hope that it will help him to better understand how to evaluate problems and correct them. This book should provide information that will help departments and individuals to select the ammunition, and even handguns, that will do the job far better than existing equipment. We cannot separate the bullet from the ammunition, or the ammunition from the handgun; they are all wrapped up together into one unit. Therefore, we must touch on all angles that relate to ballistics.

CONTENTS

PRACTICAL
HANDGUN BALLISTICS

PART I
Interior Ballistics

Interior ballistics deals with what happens from the time the handgun firing pin strikes the primer metal to the time the projectile leaves the handgun muzzle and enters the air. Many things take place but nothing can change one single fact — the handgun is only a projectile launcher. All the handgun does is guide the bullet to the point of bullet impact. It does not, and cannot, affect or alter what the bullet does at the target.

However, the handgun can have a long barrel or a short barrel, thus slightly increasing or decreasing the expected bullet velocity with resulting change in what the bullet may do at the target. The handgun can have maximum tolerances or minimum tolerances. It can have a tight bore or an oversize bore. While these details can all affect, in some manner, what the bullet does at the target, basically, the handgun remains merely a projectile launcher just as a mortar barrel directs the flight of the mortar shell but has nothing to do with what happens when the mortar shell strikes the target.

Unfortunately, there are few things in ballistics that are always true. We do not have many constants and without constants, we cannot make definite, specific statements. I have talked with attorneys and quite often they become irritated because I answer questions with "yes, that's true providing . . ." or "right, that would work if . . .", and so forth. I have had top police brass pound on pamphlets and books and tell me that such and such is true because "it says so here." My reply is simple — "under those conditions what they tell you is right, but those conditions do not apply in 97 percent of the cases in your city/area". This is the fascination of ballistics. Each situation must be taken as an individual case. Even the handgun tolerances can, at times, affect the results or so influence deductions as to make them questionable. Barrel lengths have often contributed directly to the continued existence or demise of an officer merely by increasing or decreasing bullet velocity.

You should now realize that ballistics is not a cut and dried subject; rather, it becomes an extremely fascinating and challenging one. It is necessary to retain an open mind. Psychology plays a part in ballistics, and it must be correctly evaluated in order to reach logical conclusions. Let me give you a typical example. An officer that I knew had just made sergeant and had bought one of the heavy barrel .38 Special Smith and Wesson revolvers with a four-inch barrel. It was a nice piece. We shot it on my ranges, discussed it, and shot it some more. A couple of weeks later, he became involved in a narcotics raid. Two men attacked him simultaneously. He put both of them down. When he told me about it later, he said "Man, that was a close one. They both came through that door at me and if it hadn't been for that real heavy barrel," he patted the barrel, "they would have gotten me. It sure shoots hard, man, real hard." I let him go through the routine, then asked him where he had hit each man. He pulled out pictures showing the two men lying on the floor with a single bullet hole between their eyes. "Two shots, man," he beamed, "my old light barrel would never have done this." I then broke the news that it wasn't the heavy barrel but his highly trained shooting reflexes that had brought down the two attackers. He still needed convincing, so I set up my chronograph and proved it to him. This is the type of reaction that must be evaluated in order to arrive at the facts of the matter when dealing with ballistics. Barrel weight did not affect what the bullet did at the target.

Chapter 1

THE CARTRIDGE

CARTRIDGE components are an integral part of ballistics. For some reason, they are misunderstood by the average person, so I believe that we can only get into this subject by examining each component. There is no such thing as "the most important component." Leave out one component and you do not have a cartridge. It is true that you can use cheap components but you must have a cartridge of some kind.

First, nomenclature is vital. We do not speak of a bullet and mean a cartridge. We do not mention a cap and refer to a primer. We do not fire a bullet; we do fire a cartridge.

The cartridge consists of a brass case, a primer, powder, and a bullet. It is the cartridge that is inserted into the chamber and fired.

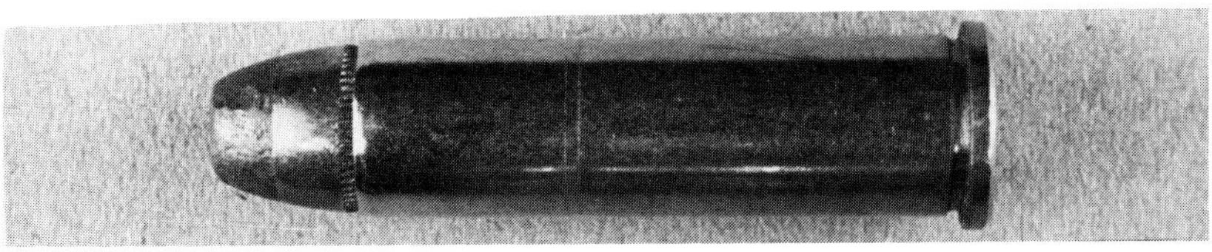

Figure 1-1A. A cartridge.

Figure 1-1B. Cartridge components: the bullet, the powder, the cartridge case, and the primer.

THE CARTRIDGE CASE

Once the cartridge has been fired, the fired case remains —

together with the fired primer — in the primer pocket. This is not a shell; it is a cartridge case. This cartridge case with its fired primer may then be removed from the chamber of the gun and either discarded or retained to be reloaded and fired again.

This cartridge case was produced as follows. Strips of sheet brass are run through a machine that punches out discs of predetermined size. These discs are then cleaned and run through a series of machines that cup and draw these discs to the desired length and size. Bases are formed, heads created, primer pockets punched into the case heads, flash holes either drilled or punched. Finally, the case length is checked and trimmed. It is a simple, but very noisy, operation.

There are two basic types of handgun cases: rimmed and rimless. Revolvers require rimmed cases, although Ruger produces a revolver that chambers and ejects the .30 caliber carbine cartridge using springs and cams fitted into the extractor head. Automatic pistol cases have rimless heads. Automatic pistols have been modified to use rimmed .38 Special cartridges, but generally, pistols fire rimless cases.

In the old days, it was far easier to punch out and create the case head and primer pocket by bending the brass to produce balloon head cases. Black powder pressures were low and this proved to be an acceptable and safe method of forming case heads. Today, case heads are solid brass and extremely thick to safely contain the pressures developed by modern handgun cartridge loadings.

Primer pockets come in two *sizes*: large and small. The small pockets take the 0.175 primers and the large pockets the 0.210 primers. You can still find .38 Special cases with large primer pockets. Many .45 ACP cases have undersize primer pockets to take Special primers developed by the government many years ago. The novice should never become upset when he runs into something unusual in the firearms and ballistic field. Just about everything has been tried at one time or another. Strangely, many of these oddball details continue to circulate today.

There are also two *types* of primer pockets. The Berdan primer pockets have a solid nipple jutting up from the bottom of the primer pocket that acts as a built-in anvil; on either side are two small flash holes through which the primer flame passes. The second type is the Boxer primer pocket that has only a single round flash hole in the bottom of the primer pocket.

Functionally, there is little or no difference between the two types. From a reloading point of view, the Berdan type is extremely slow and difficult to work with. The most common is the Boxer and is universally used today throughout the country, although the Berdan remains popular in Europe.

The primer is a small, metal cup created out of specially selected metal to meet the requirements of the average cartridge. This cup is partially filled with a primer compound that, when dry, can ignite and produce a flame. An anvil is inserted within the cup. The anvil is a three-legged cone with the high (top) side touching the primer compound in the base of the cup; the three legs rest on the bottom edges of the open base of the cup. When correctly seated into the primer pocket of the cartridge case, the three legs touch solidly upon the bottom of the primer pocket, presenting a firm base for the primer. Thus, the primer compound remains in position over the top of the cup. Then, when the firing pin strikes the center of the primer cup, it dents the primer cup metal and crushes the primer compound between the cup metal and the top of the anvil; if all has gone well, the primer compound fires, driving a long cone of fire into the inside of the cartridge case.

Each manufacturer has its own primer compound·formula so that no two primers from different manufacturers are exactly the same. The force of the flame, its intensity, and its duration vary from manufacturer to manufacturer. There are primers created for small pistol cartridges, large pistol cartridges, small rifle cartridges, large rifle cartridges, and also for shotshells. Each type has its own characteristics. They should never be used for any cartridge for which they were not designed. Primers are literally miniature hand grenades and are extremely dangerous if mishandled or misused.

In recent years, another type of primer has been brought out. This is the Magnum primer designed for use with slow-burning powders in relatively large cartridge cases. It has no relation to Magnum cartridges; Magnum cartridges may be loaded with standard primers. However, the use of Magnum primers in Magnum cartridges is often advisable when slow-burning powders and heavy bullets are also used. Magnum primers have entirely different burning characteristics than standard primers. As Scott Heter of CCI once told me, "Standard primers are like gasoline. Toss in a match and they go blam! Magnum primers are like kerosene, producing a long, soft flame that, instead of shoving around the powder, will softly ignite the powder over a period of time." I have always thought this was an excellent description.

Primer cups are stamped out of strip metal, then the discs are cupped, trimmed, cleaned, and readied for filling. Because handguns have a lighter firing pin fall than rifles or shotguns, the metal used in handgun primers is thinner and softer than that found in rifle primers. Rifle and handgun primer cups come in the same sizes, large and small. Outside dimensions are critical because these cups must fit snugly into primer pockets, yet must not be too large.

The primer compound is a very touchy substance and is usually mixed wet. It is then pressed into the bottom of the primer cup and often covered with a sheet of foil or special paper. The anvil is then set into the cup, the top of the cone faces up, not quite touching the primer compound, and the three legs are firmly locked into the open end of the primer cup.

Primers are simple and yet more time, money, and control are lavished upon them than any other component part. Primers are extremely dangerous and should always be handled with great care. Do not attempt to pry the anvil out of a primer to show it to someone; also, do not scrape out the primer compound. Once this compound dries and is set, it is ready to fire. If it is absolutely necessary to remove the anvil or primer compound for some reason, soak the primer overnight in lubricating oil, then wearing glasses and gloves, use great care in prying the anvil out of the primer cup. There should never be any doubt as to the potentials of any primer.

Use care with powders, but with the exception of black powder, which can detonate for no good reason at the most embarrassing times, one can work with modern day production powders without any great danger — provided they are confined only within the original containers. Years ago, in New York State, I had to build a powder magazine. I had always kept my powders on open shelves; if a fire occurred, the powders would only pop the light containers and burn, there would be no detonations. I was told to place my powders within a heavy metal and wood container, because they "would then be safe." No amount of discussion could convince the authorities that they were ordering me to make a homemade bomb. If, for any reason, my powders had ever gone off within the mass of metal, it would have put to shame a three-inch shell. Powder should be kept within a very flimsy container far from flames, heat, primers, and other potential sources of dangerous materials. An old refrigerator is an excellent storage place for powders because they are protected from heat and flame and yet, if they do let go, they will merely burn and not detonate.

I continue to remain highly amused at the airline personnel who get panicky when they see a few cartridges lying around in my brief case. These are usually confiscated and I am then watched very carefully. I don't know when this attitude started, but it is of long standing. Throw a cartridge all by itself into a fire and it will go POP! A cartridge must be contained before it will fire in the normal manner, which is why there are chambers and recoil plates and breech blocks to securely hold the cartridge.

I once knew a man who would don a pair of asbestos gloves, hold a .38 Special cartridge between his fingers, and then hold a blow torch on the primer. When the primer fired, all that happened was that the bullet fell on the floor eight or ten feet away; the primer usually blew part way out of the primer pocket and his fingers remained unharmed. During the war, I watched men amuse themselves throwing bandoliers of machine gun ammunition into fires. Unless the case is confined, powder gases cannot build up sufficient pressure to produce normal firing specification. True, every now and then a case will rupture and that is why my friend wore asbestos gloves; as for being "dangerous," they are not.

Pour some powder onto a piece of concrete, then touch a match to it. Watch it burn slowly, intensely, and progressively. Never attempt this with black powder; black powder will detonate. Try various types and makes of smokeless powder and watch their burning rates. Learn how to work with powders. Respect them, but do not be afraid of them. If, at any time, you note a light brownish dust appear, or the powder grains seem to be disintegrating, get rid of the powder. It is decomposing and can become erratic in its performance. Normally, modern powders (by Hercules, DuPont, Olin) will last for years if kept in a cool, dry, clean place in their original containers.

We have three basic powders today: disc, ball, and granular. Look-alike powders are not necessarily the same or even close to having the same burning characteristics. A person should not get up on the witness stand and testify that "this is Unique powder because I can tell by the powder grains"; he should have a professional lab analyze the powder.

Handgun powders are a varied and assorted lot with tremendously varying burning rates and powder grain shapes. Rifle and shotgun reloaders will note that many of their powders are also excellent handgun powders for specific purposes. There is no one ideal handgun powder. Both the handloader and the ammunition companies must use the powder that appears to best do the job at hand. Powders are blended at the manufacturing plant so that each lot will perform within prescribed limits. When ammunition companies order powder, they usually develop a load from the specific lot that will do the required job. As an example, they may order a powder that is basically Unique; however, tests show that it is, perhaps, slightly faster burning than canister Unique sold on the market to handloaders. The ammunition company will then conduct its own tests to determine what loading should be used. For this reason, it is impossible to pull bullets from factory ammunition and make any intelligent statement about loading density, type of powder, etc.

THE PRIMER

We must now return to the beginning of the firing sequence.

In order to fire a handgun, somehow the firing pin must be struck by a hammer, or released to fall under its own spring tension to strike and fire the primer. Some mechanisms require the bolt to fall, rammed forward by a heavy spring to force a nipple, machined into the face of the bolt, to hit the primer and fire it. In any event, once the trigger has been moved to the point where it releases a mechanism that must move to actuate the firing pin, stresses and vibrations are instantly set up within the handgun. Remember that the primer has not yet been struck. I have watched these tests in factories. The length of time that it takes for the released hammer or firing pin to commence its fall right on through to the time the firing pin actually ignites the primer is known as lock time. The shorter the lock time, the faster the ignition. In precision handgun and rifle shooting, this lock time can be vitally important.

Figure 1-2A. The firing pin point strikes the primer and fires the cartridge. (Smith and Wesson revolver.)

Stresses and vibrations are set up and move throughout the handgun. The moment the firing pin hits the primer, additional stresses and vibrations commence. Few people appreciate

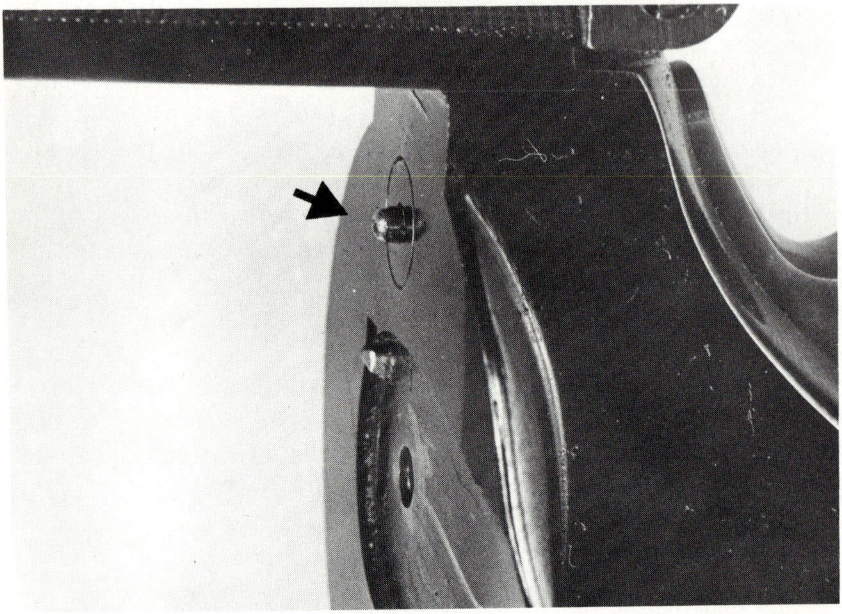

Figure 1-2B. The firing pin point or tip protrudes through the recoil plate of the revolver frame to strike the primer.

the potency of the primer. As a demonstration, load a cartridge case with only a primer — no powder or bullet. Turn off the lights, hold the handgun in front of you, and pull the trigger. Most people are awed by the pyrotechnic display put on by the lowly primer. Don't ever sell the primer short: a box of one hundred large rifle primers, correctly placed and detonated, can do some amazing things.

Figure 1-3. A Colt four-inch barrel revolver firing a .357 Magnum primer in an empty cartridge case. The flash is visible (left) in the gap between the front of the cylinder and the rear of the barrel throat. Note the amount and intensity of the flame as it bursts from the muzzle (right).

Most standard primers of the same type will produce similar results so that, by the time the primer flame has passed through the flash hole into the powder chamber, things begin happening, both to the handgun mechanism and to the cartridge. Up to this time, the cartridge has been a stable unit; it now begins to move back within the chamber. The primer itself will tend to move backward and out of the cartridge primer pocket, because it is far lighter in weight than the powder and bullet that lie ahead of it. Pressures on the primer come from in front of the primer, thus the tendency to move back and out of the primer pocket.

Usually, the cartridge case is fairly well filled with powder so that the primer flame is directed upon a mass of powder that will ignite at the rear and, as pressures build up, force the bullet forward out of the case mouth. As more and more powder burns, the space between the rear of the cartridge case and the base of the bullet increases, thus keeping down pressures and maintaining a steady mass of pressure against the bullet base.

This is an excellent place to discuss the 2.7 Bullseye Factor. During the past forty years, I have seen four detonations, three occurring with .38 Special revolvers and one with a .357 Magnum revolver. At the time, and even as recently as early spring of 1977, I chalked them off to carelessness in handloading or abusing the Star Machine. Since 1955, I have also been aware of problems with #4831 rifle powder. This rifle powder is very slow burning and superbly adapted to large capacity cases and heavy bullets. I have personally seen three rifles blown up by this powder

When we evaluate the traditional handgun target loadings of 2.5 to 3.0 grains of Bullseye with a standard small pistol primer, the 146 grain cast wadcutter bullet in the .38 Special cartridge, it appears to be both remarkably safe and superbly accurate. Why then is there any problem with this load and what is the problem? The problem is simply that this loading can detonate. It has only been recently that any attempt to seriously investigate these occurrences has taken place, due to the efforts of Jeff Cooper to obtain as much information about this as possible from all over the country. He did not pass off these detonations as human errors the way I did. The more I think about his comments, the more I believe they have merit. I do not believe that anyone can yet make any specific statements of fact. So far, all is surmise, but it is a situation that should be given thought.

Bullseye powder is a fast-burning powder. As a result, only small amounts are used. The .38 Special case is a black powder case and far too large for efficient powder burning potentials with any but the slowest burning powders. This indicates that under certain conditions, when the handgun is about to be fired, the powder can lie on the bottom of the case. This thin layer covers the bottom of the case from the rear to bullet base. If conditions are just right, when the primer flame passes through the flash hole, it does not directly strike the powder, or mass of powder, but continues through the flash hole to the bullet base *above* the powder; thus spraying the entire powder charge with flame all at one time. There is, therefore, no progressive burning and the powder charge detonates, blowing up the revolver.

There are reasons to believe something like this must take place. When one considers the many billions of Match Target .38 Specials loaded with from 2.5 to 3.0 grains of Bullseye that have been fired in the last forty years, the percentage in favor of a detonation is almost nil — yet, it does happen. It is possible that certain lots of powder lend themselves to a detonation; however, I have watched powders created, manufactured, blended, and mixed and then watched the testing and evaluation of every lot. It is difficult to imagine that the fault lies with the powder, so I am inclined to favor the explanation given above.

As for #4831 powder, let me explain that full charges have never created problems. If a full powder loading requires 71.4 grains of #4831 powder and if charges varying from 65.0 to 73.0 grains are used, there has never been any trouble. Problems arise when a handloader tries to use a reduced charge of perhaps 51.0 grains of #4831; not always, yet frequently enough to indicate that greatly reduced charges can cause detonations. It would appear that these detonations occur for the same reasons as with the .38 Special and .357 Magnum detonations:

powder lying on the bottom of the cartridge case and the primer flame igniting all of it instantly, rather than progressively.

Similar detonations have never, to my knowledge, occurred with any automatic pistol cartridge. This is probably due to the correct relationship between automatic pistol cartridge case dimensions and the powders used in these cases. Pistol cases are considerably smaller than the large black powder cases used for revolver ammunition and are thus far more efficient.

We must now turn to the powder charge and examine what happens when it commences to burn.

THE POWDER

Powders are very individualistic; each has its own very special burning characteristics, which may lend themselves to a wide range of loading potentials or, conversely, to an extremely limited loading potential. Some powders may be used for shotshell, handgun, and rifle cartridges, within limits of course. Others are not at all versatile. Some powders that look precisely like other powders will have greatly differing burning characteristics. Therefore, it is extremely dangerous to attempt to guess what a specific powder may be.

Great care is taken by the powder manufacturers to make certain that each basic powder remains the same, retaining its burning characteristics over a period of many years. Tests are constantly being run so that a constant uniformity exists. Because of this care and control, powders usually perform precisely as expected.

Obviously, fast-burning powders will be used in smaller quantities than slow-burning powders. As an example, assume that Bullseye powder will be used in a .38 Special cartridge. This powder will peak or reach its maximum pressure far faster than a slower burning powder like #2400. When powder burns, it creates heat; the higher the heat, the more erosion will occur in the chamber and barrel. On the other hand, slow-burning powders often cannot be used due to their inability to burn completely behind the selected bullet weight, nor can they always be counted upon to deliver the desired bullet velocities.

Conversely, fast-burning powders may peak too rapidly, creating pressures that will strain the handgun; also, their burning time is too fast to allow the bullet to reach the desired velocity.

As a result, quite a bit of thought must be put into choosing the correct powder for the job at hand. Not only must the powder produce the desired bullet velocity but it must do so by producing pressures that fall within accepted limits. There should be little or no unburned powder left within the chamber and barrel after firing not just one but hundreds of rounds.

Figure 1-4. The powder grains on the left have already burned; only a residue remains. This powder string was about three inches long; it took nearly seven seconds for the entire string to burn. Unconfined smokeless powder is usually not explosive.

What happens when the primer flame passes through the flash hole and strikes the powder? Initially, a small amount of powder commences to burn. Today, progressive-burning powders are coated with substances that slow down or delay the burning rate of the exterior skin of the powder particles so that, at the start, powder combustion is slowed; as the skin burns off, the burning rate increases. By that time, pressures have built up so that the bullet is moving, thus rapidly increasing the combustion area within which the powder will burn. Because of this, pressures cannot become excessive and continue to remain within acceptable limits. Due to the ever-increasing burning rate of the powder, pressures usually remain behind the bullet base to provide a steady push to the bullet, thus ramming it out of the handgun muzzle at the desired velocity.

Without this progressive-burning characteristic, powders would detonate, and it would be difficult, if not impossible, to obtain high bullet velocities.

As the powder commences to burn, pressures immediately build up, forcing the cartridge case walls outward and backward. Now that the cartridge case is holding the ever-increasing pressures of the hot gases, the walls expand and the case moves back slightly to force the primer back into the primer pocket and lock the cartridge case solidly into the chamber If there were no release for the gases, the handgun would explode but since the bullet is moving, the pressures are relieved by providing an ever-increasing area between the rear of the inside of the case and the base of the bullet. Once maximum pressures have been reached, the cartridge case brass springs back down slightly. If this did not happen, it would be difficult to extract the fired case from the chamber.

As a result of the pressure peak and the relative softness of the cartridge case brass, at this instant the case is marked by specific characteristics of the handgun chamber and breech face, along with the identifying marks of the firing pin in the primer. This is why it is possible to match a fired case to the specific handgun that fired it, provided, that no attempt has been made to alter the breech face, firing pin, or chamber characteristics.

When the hot gases strike the base of the bullet that is securely held within the case mouth, the bullet tends to resist the force of the gases. Most of the time, the bullet will expand first as far as possible within the confines of the cartridge case mouth, then out to the maximum diameter of the chamber. Next, the force of the gases will break the bullet loose of the case mouth, and the swelling will continue throughout the entire bullet, limited by the size of the chamber. The hotter the loading, the more the bullet will be deformed at the start.

As a demonstration, remove the barrel from a revolver and then fire several high-velocity loadings out of the cylinder, through the front of the frame, into oiled waste. The recovered bullets tell their own story.

It is possible to actually melt the base of a lead bullet, thus adding to bullet deformation; this could explain why many loadings that appear to be excellent on paper just do not work out in practice. If the bullet does not remain true, accuracy will be poor.

In order to better understand this, we must discuss the differences between revolver and pistol mechanisms (see Chapter 2).

The following lists some of the more popular handgun powders that are sold in canister lots throughout the country. These powders are sold to handloaders and their specifications (burning rates, loading data, appearance, and other details) are very closely monitored at the plant. The list is in order of the burning characteristics, from slowest to fastest:

1. DuPont 4227
2. DuPont 4759
3. Olin 296
4. H 110
5. Hercules 2400
6. Olin 630
7. Hercules Blue Dot
8. Herco
9. Unique
10. Green Dot
11. Hi Skor 700X
12. Red Dot
13. Olin 230
14. Bullseye

Burning Characteristics of Powders

Powders are very individualistic and usually have extremely critical burning characteristics. One powder may be ideal for low-velocity loadings with wadcutter bullets and, when used in small quantities, can be ideal for the specific purpose. Unfortunately, Bullseye is being critically examined at this time to determine whether it contributes to detonation under certain conditions. It is customary to use 2.5 to 3.0 grains of Bullseye with the wadcutter 146 grain bullet. This makes an excellent match loading that, with the correct bullet, can provide the necessary accuracy.

There have been times when such a load has detonated and wrecked .38 Special revolvers. I have had this happen to me three times over the years. I have known of other instances, but they have not been frequent enough to warrant an in-depth evaluation. Recently, Jeff Cooper has brought this situation to the attention of match shooters in an attempt to determine whether or not the powder is at fault. His theory, and it does appear quite reasonable, is that under certain conditions the powder charge can, and does, lie flat upon the bottom of the cartridge case. There is so little powder in such a large capacity case that there is no question but what it can happen. Then, when the primer fires, it ignites not just a small portion of the powder but the entire powder charge at one time, thus creating a detonation with resulting excessive pressures. Remember that Bullseye is a fast-burning powder.

This is merely one small phase of the powder burning situation. While the chances are remote, such a detonation can occur, and it points up the thought and care that must be put into powder evaluation. Far too many people believe that progressive-burning powders are extremely safe. They are, *but* their individual characteristics must be examined in depth.

Another example of how touchy some powders can become under certain conditions is Herco, an excellent and widely used powder. On the other hand, I have discovered that it too can become violent. As an example, let us assume that 10.0 grains in the .44 Magnum is a fine, maximum, safeloading. It is possible to load from 9.0 to 10.0 grains with a specific bullet and obtain excellent results. Suppose that the handloader decides to take his revolver and go hunting with it. Because of the clean burning characteristics of Herco, he wishes to use it; he also wishes to slightly increase velocities, he uses 10.8 grains, or even 11.0 grains, of Herco. At his point, Herco can become erratic, giving excessive pressures and recoil without any compensating increases in velocity or bullet performance.

Details like this run throughout the entire powder line. Each powder must be examined and evaluated, balancing the good aspects against the bad. Basically, a powder must be chosen as a part of an entire cartridge so that the whole will produce satisfactory results. Bear in mind that bringing together four components does not necessarily mean that the final cartridge will be safe or even practical, even though each component has been chosen as theoretically ideal.

A lightweight bullet should be used with a relatively fast-burning powder. A heavyweight bullet should be used with a relatively slow-burning powder. In theory, this is fine. On the other hand, higher velocities may be obtained with a lightweight bullet and a case full of slow-burning powder. Why then not use a slow-burning powder behind a lightweight bullet? When a powder does not burn completely, it leaves residue within the case, within the chamber, and within the barrel. The hot gases also blow the residue into cracks within the revolver mechanism. Quite often, less than a dozen rounds need be fired until the revolver mechanism locks up due to accumulation of powder residue. Obviously, this is not desirable.

It is desirable to load as heavy a powder charge as possible so that the interior of the cartridge is filled with powder. Ignition is more uniform and velocities tend to become more consistent. Because of the powder residue, it is more practical to use a faster burning powder and to lose a small amount of velocity, thus ensuring the correct functioning of the revolver over long periods of firing.

Slow-burning powders require compression to burn correctly; therefore, they are customarily used behind heavy bullets so that pressures remain within accepted limits and powder residue is held to a minimum. A favorite powder for maximum loads in the big Magnums is #2400. Unfortunately, #2400 leaves substantial residue, and the handloader must weigh the advantages against the disadvantages. An excellent Magnum powder is #4756; it burns clean under most conditions, yet produces velocities commensurate with #2400 without any of the disadvantages.

As may be seen, all powders have advantages and disadvantages. I regret that factories continue to use powders that leave substantial residue, particularly in the full power Magnum loadings. I believe that much of this is due to the desire to maintain high velocities — in other words, a competitive merchandising situation exists. While perfectly safe, these loads can create a great deal of residue.

The ideal powder would be one that would just about fill the case, burn relatively slowly, leave little to no residue, and provide the required velocity with pressures within acceptable limits. Such a powder does not exist. We are forced to compromise throughout the entire range of powders in an attempt to obtain the most satisfactory results.

THE BULLET

The bullet is the primary component in the cartridge because it is the only one that reaches the target. Bullets present an almost unlimited field of potentials. It would be nearly impossible to delve into all their variations over the ages. Because of this, I want to stick closely to current bullets, considering only the single one-piece projectile.

Generally speaking, all bullets are sized by the manufacturer to fit into the intended barrel. They will usually fill out when fired; they then fit into the grooves cut into the barrel, thus permitting the raised barrel ridges, or lands as they are called, to rotate the bullet as it passes down through the barrel. Soft lead bullets tend to skid because the soft lead does not allow the lands to grab the bullet and instantly rotate it. Hard lead bullets let the lands grab earlier and faster, but hard lead bullets may be pushed at higher velocities because the bullet will not deform as rapidly as a soft lead bullet. Jacketed bullets must have a jacket that will "take" the rifling of the barrel without ripping it to pieces.

Lead Bullets

Cast Lead Bullets

In order to obtain a cast lead bullet, lead of some kind must be melted in some manner, and then a small amount of this lead poured into a mold, allowed to cool and harden, and then removed from the mold. If properly cast, this bullet then closely resembles its final shape. It will usually have a grease groove or two and often be slightly out of round. It should have no interior hollows, cavities, or other similar defects. It is what we call an unsized, unlubricated bullet. It is the first step in producing a bullet.

The rough, unlubricated, unsized bullet can be fired from a

handgun precisely the way it comes from the mold; it is not impossible but is impractical. The ballistician should always keep this in mind. Bullets fired in this condition often are impossible to relate to a specific barrel due to barrel leading.

Cast bullets generally fall into several categories, of which soft lead and hard lead are the usual descriptions. They may be cast from pure lead ingots, from type metal, or from tin alloy or a combination of alloys. Quite often, it is relatively simple to tie a bullet composition to the raw alloy at hand in a suspect's shop.

Swaged Lead Bullets

For the past few years, we have had an increasing number of commercially swaged lead bullets for match shooters and police practice ammunition. These bullets have lubricating grooves that are generally shallow. Rather than using conventional grease lubricants, or even the new synthetic lubricant, these new bullets are run through a liquid solution; as they emerge, the liquid rapidly dries leaving a thin coating of a waxlike lubricant all over the bullet. This lubricant is clean. The bullets are easy to handle and do not usually clog the seating die of loading machines like conventional grease lubricants. Leading is held to a minimum at normal match velocities.

Gas Check Bullets

Lead bullets have been widely used for hundreds of years and, even today, far outnumber jacketed bullets because lead is relatively cheap, is easy to find, and may be alloyed to produce a bullet of the suitable hardness for the job at hand. Handgun hunters learned many years ago that it is difficult to push lead bullets at high velocities due to resulting extreme leading of the barrels. Leading basically occurs when the hot searing gases of the propellant powder slam against the exposed lead of the bullet base. Lead melts at relatively low temperatures, and the bullet bases become deformed. Lead bullets become very hot during the passage from the cartridge mouth through the barrel. In addition to a partially melted bullet base, bullet surfaces that come in contact with the barrel can leave films of hot lead within the barrel. Ordinary, straight lead bullets cannot usually withstand the heat of high velocity loadings.

Because the bullet base is vital to accuracy, and because it is directly in contact with the heat of the powder gases, shooters learned many years ago that a copper cup affixed to the bullet base would help protect the bullet base and allow much higher velocities. These cups are known as gas checks and usually are fastened to the bullet base during the bullet sizing and lubricating process. Gas check bullets have been used for hunting purposes for many years.

Why Must Lead Bullets Be Sized and Lubricated?

Cast lead bullets are usually formed in molds that are slightly oversize. This permits the true shaping and final forming of the bullet as it is run through a sizing die. There are many types, but the basic operation is the same. The die is precisely the correct size for the barrel of the handgun in which the bullet will be fired. As an example, .38 revolver bullets are normally size 0.357 whereas 9 mm. pistol bullets are sized 0.355. When the bullet is removed from the die, it is now perfect, true, and ready to be lubricated.

Figure 1-5. Cast lead bullets as they come from the bullet mold.

We cannot, from a practical point of view, fire an unlubricated bullet. Lead scrapings from the bullet, plus a certain amount of deformation from unprotected base bullets, will adhere to the inside of the barrel, sticking to the lands and grooves. As additional bullets are fired, this layer of erratically formed lead grows until finally it can coat the barrel from end to end. The softer the bullet lead, the faster the accumulation of

Figure 1-6. Cast lead bullets after they have been sized and lubricated using a conventional heavy grease.

Figure 1-7. Cast bullets lubricated with a white, dry, waxlike lubricant.

barrel lead.

In order to prevent this lead buildup, lubricating or grease grooves are formed into the bullet. There may be one or two, or even more, around the bullet circumference. The depth of these grooves varies. When the bullet is run into the sizing die, most setups also lubricate the bullet — that is, force lubricant into the grease grooves of the bullet. Thus, the bullet is both sized and lubricated in a single operation. This lubricant tends to prevent lead buildup within the barrel and also aids in lengthening the barrel life.

The type and composition of the lubricant is determined by bullet velocity. Some lubricants are excellent for low and moderate velocity loadings. Others allow much higher velocities without leading. The barrel surface does have much to do with leading. There are many barrels in existence today that will lead regardless of what type of lubricant is used or how hard the bullet lead.

Lubricants are a necessity but they do have disadvantages. The bullet moves out of the case mouth under increasingly hot blasts of powder gas. This heat is transferred to the bullet and to the lubricant. By the time the bullet reaches the end of the cylinder and jumps the gap into the barrel throat, part of the lubricant begins to vaporize. This tiny cloud of vaporized lubricant has been blasted out of the front of the cylinder under high pressures. The moment some of it strikes the cold metal of the revolver, it solidifies. Some of it, driven by the high pressures, finds tiny gaps and holes and crevices. After a few shots of certain loadings, the lubricant builds up. Within perhaps a dozen rounds, the revolver cannot be opened or closed due to the buildup of the vaporized lubricant between the crane and the frame — under the extractor, around the front of the cylinder, along with tiny particles of lead that have bounced off the rear of the barrel. Other deposits will grow on the rear of the barrel itself. Generally, it takes high-velocity loads to accomplish this in a few rounds. Much depends upon the lubricant itself and its composition. For this reason, revolvers using lead lubricated bullets must be cleaned constantly.

Despite cleaning, most people fail to completely remove all of the blasted, vaporized lubricant from all of the cracks and crevices of the revolver mechanism, thus permitting a fairly easy match of lubricant on a pistol to bullet lubricant.

FIRING UNLUBRICATED BULLETS. Many years ago, I loaded up a couple of dozen match wadcutter bullets that had not been sized or lubricated. Velocities ran around 750 fps. Accuracy was surprisingly good for a dozen rounds despite the rapid leading of the barrel. Accuracy fell off around the fifteenth shot. By that time, the barrel was pretty well coated with lead. I continued firing out of curiosity. Lead buildup became so bad that I never did finish firing all the cartridges. This does not prove anything terribly important. All it does confirm is what most experienced shooters assumed would happen. On the other hand, it would have been nearly impossible to match a fired bullet to the barrel after perhaps a dozen shots.

Bullet Base

One of the most important accuracy requirements is a perfect barrel muzzle configuration. If more gas blows out one side of a bullet base than another, that specific bullet will be blasted slightly off course. As a result, it is imperative that the bullet base be perfectly true, round, and flat before the bullet is fired and remain that way until the bullet reaches the target. Any erosion of the bullet base will affect bullet accuracy because of the slight alteration of the surfaces upon which the gases are pushing as the bullet leaves the muzzle of the handgun. Any erosion on the circumference of the bullet base will tend to tip the bullet as it leaves the muzzle due to uneven gas pressures at the last moment. Gas checks help considerably in keeping the bullet base true. They also help keep down leading by a lead wiping effect as they pass through the barrel.

The bevel-base bullet is widely used because of its ability to feed easily through automatic loading machines. For some unknown reason, this type of bullet base is not conducive to accuracy. When fired in comparative tests against hollow-base and flat-base bullets of the same type, the bevel-base bullet will

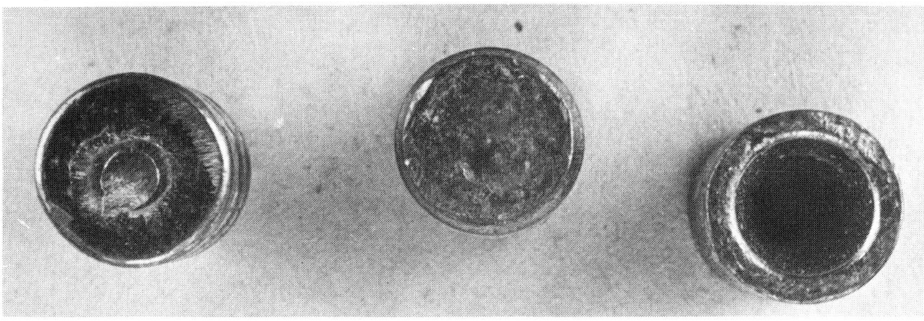

Figure 1-8. (Left) a cast bullet base, (center) a jacketed bullet base, and (right) a swaged, lead, hollow-base bullet base.

not provide similar accuracy.

The hollow-base bullet will generally produce better accuracy than either of the other two types. This concept goes back to the theory that if the bullet base is hollow for a considerable distance of the bullet body, the pressures of the hot gases will force the bullet base to expand, forcing the skirts outward against the walls of the cylinder chambers and the barrel grooves, thus effectively sealing the barrel against any gas leak. The problem is that hollow-base bullets cannot be pushed at high velocities. In order to effectively seal the barrel, the hollow-base bullet skirts must be relatively thin; under the higher pressures and ever-increasing torque forced upon the bullet by the barrel rifling, the skirts have a tendency to break off, thus destroying any potential bullet accuracy. They are, however, excellent low-velocity match bullets.

The flat-base bullets are perhaps the best all-around bullets. They appear to have fewer problems than either of the other two types. On the other hand, the flat base must be true, clean, and absolutely perfect all the way around.

As may be seen, the lead, cast, or swaged bullet has many advantages, the main one being its low cost. As long as velocities are held to bullet tolerances, they are a very practical projectile.

Bullet Nose Shape

There are basically four types of lead bullet noses. First is the wadcutter commonly used for practice, target, and plinking. The flat nose type chops nice clean holes through target paper. It is not a high velocity bullet nor is it accurate at long ranges.

Second is the hollow-point bullet that is a hunting type bullet. This design forces the bullet nose to expand shortly after impact, depending upon the softness of the bullet lead and the depth and size of the hollow point.

Third is the round nose bullet that is traditional but is not generally accepted today because of its ability to penetrate excessively and deflect upon striking hard surfaces. It was originally (1880s) thought that the round nose would enable the bullet to fly more precisely to the target, but today this thinking is obsolete.

The fourth type of bullet nose is the Keith type. This has a large flat nose. The base of the nose rests upon the body of the bullet with a shoulder around the entire base of the nose. When this bullet strikes, the nose tends to give way and expand but the heavy, solid shoulder chops its way through heavy muscle and bone to help drive the bullet deep into the game animal. It is an excellent bullet design and one that has retained its popularity.

Today, we have other nose designs, including a new one with a deep cavity and a lead rod that comes up through the bottom of the nose cavity. We will continue having new concepts as long as we have lead bullets.

Jacketed Bullets

Jacketed bullets are a relatively new development in the handgun business, which is strange because we have had jacketed rifle bullets since the turn of the century. For some reason, there was no commercial production of jacketed handgun bullets until the 1950s and that was sparked by Jim Harvey of Lakeville Arms. Then Lee Jurras brought out his Super Vel bullets, and everyone started turning out variations on jacketed handgun bullets. The concept is simple. Encase pure lead within a gliding metal jacket and you can push this projectile at extremely high handgun velocities.

The execution of this concept was far more difficult than

many people imagined. Short, half-jackets were quite inexpensive and they allowed velocities to increase. Unfortunately, if velocities were not high enough, the short half-jacket had a tendency to pull off the bullet during passage down the barrel. The next bullet would strike the stuck jacket and problems ensued. Longer jackets were tried with varying degrees of success. Warnings were put out to keep bullet velocities high enough to prevent jackets from coming off the bullet bodies. Quite often, the jacket would stay on through the barrel passage, but then fly off the bullet in the air.

Figure 1-9. Half-jacketed bullets.

Jim Harvey solved this problem by crimping the jacket onto the bullet body with a heavy crimp. This not only locked the jacket onto the bullet but also decreased the bearing surface of the bullet, permitting even higher velocities. The shorter the bullet-bearing surface, the less drag, less heat, and less friction there is between the bullet and the surfaces of the barrel. This means lower pressures, with resulting ability to use more powder and increase velocities.

Most of the early jacketed bullets retained the Keith type design that produced very high impact shock when the bullet struck the target. This, plus the use of pure lead cores, resulted in excellent bullet performance at the target. During this transition period, a great deal was learned about bullet jacket material. If the material was too soft, or too thin, the barrel rifling would often rip it to pieces. If too thick and too heavy, bullet expansion was restricted.

In addition, bullet base design presented many problems. If

too round, accuracy fell off. If too square and sharp, the bullet jacket base would often blow right out through the barrel with the lead, leaving the jacket walls tightly locked into the barrel; the next shot would usually split the barrel when the bullet struck the obstruction.

Leading was also a problem because none of the jackets covered all the exposed lead surfaces of the bullet circumference. True, the base was protected, but exposed lead was always visible ahead of the top end of the jacket. Finally, bullets were produced with very long jackets crimped over the bullet, thus eliminating much of the barrel leading.

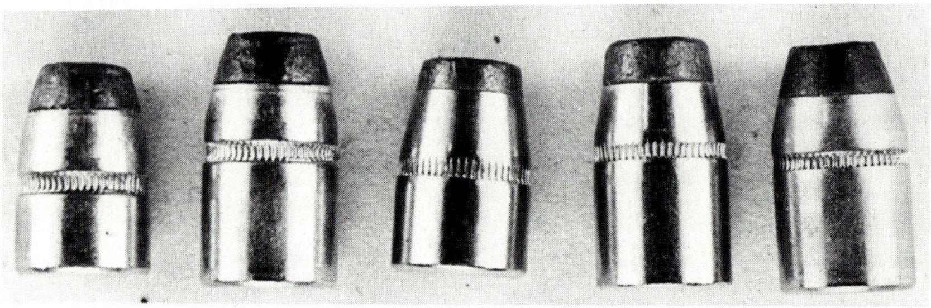

Figure 1-10. Long-jacketed bullets used primarily as revolver bullets; they are similar to rifle bullets.

In recent years, bullet design has changed drastically with handgun bullets resembling rifle bullets. Bullet-bearing surfaces have been shortened and the jackets have been carried forward over the ogive so that high-velocity bullet leading is a thing of the past. It is no longer possible to blow the lead bullet core out of the jacket, nor can the bullet base separate from the jacket walls. Most bullets utilize an exposed lead nose to aid in bullet expansion at the target. These are extremely practical revolver bullets.

Automatic pistol bullets have been even more problematic. Pistol bullets must guide the cartridge up and out of the magazine, up a feeding ramp, and then into the barrel chamber. This is done under extremely difficult conditions and the lead bullet noses become deformed during passage from the magazine into the chamber. Jams were frequent. Lead built up within the

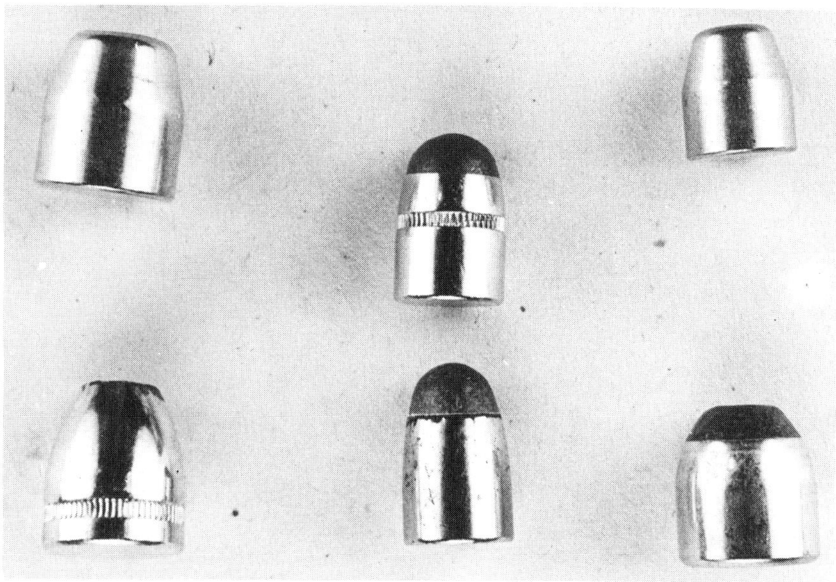

Figure 1-11. Long-jacketed bullets used primarily in automatic pistols.

Figure 1-12. Additional long-jacketed handgun bullets.

mechanism. Over the years, less and less lead was left exposed on the bullet nose until today we have automatic pistol bullets with large hollow points and no exposed lead. The bullet jackets turn over into, or completely protect, the bullet nose cavity. Therefore, during the reloading cycle of the automatic pistol, these new bullets perform as well as conventional type full-metal-jacket bullets but provide substantial bullet expansion — something that the full-metal-jacket bullets could not do.

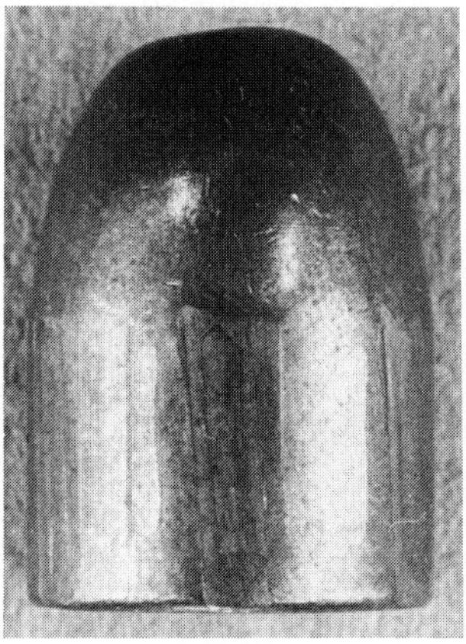

Figure 1-13. A "full metal patch" handgun bullet that has been fired; the rifling marks are clearly visible on the bullet jacket.

I have gone into quite some detail on bullets because it is the bullet that we are primarily concerned with in an examination of ballistics.

Jacketed Bullets Versus Lead Bullets

In some instances, lead bullets can equal the velocities of jacketed bullets but the requirements turn the lead bullets into a single purpose bullet. The requirements are that the correct lubrication is used, that the bullet lead is extremely hard, and that the barrel is rifled to correctly handle lead bullets. We end up with a bullet that will strike the target with relatively high remaining velocity and penetrate deeply. If a potentially fine hunting bullet for big game is the ultimate purpose of the bullet, we have obtained our objective.

Because lead bullets bear directly upon the barrel rifling, there are definite limits to what may be expected of a lead bullet. We can, of course, obtain extremely high velocities with lead bullets equal to the highest obtainable with jacketed bullets; however, after two or three shots, the barrel may become so badly leaded that it would be impractical to continue firing. Even though we can obtain these velocities, we cannot assume that accuracy will be acceptable; thus, we may have attained one goal and lost many others. Actually, it is unwise to attempt to obtain jacketed bullet performance from lead bullets. Each of these bullets had its own field. These potentials and restrictions should be respected.

Lead bullets came first. It was and remains simple to melt lead and cast it into bullets. It is an easy and cheap method of obtaining projectiles. Back in the days of black powder and cap and ball revolvers, if a handgunner wished to have a more powerful handgun, he bought one with a large caliber. In other words, a .44 caliber revolver had far more stopping power than a .36 caliber revolver. Chambers were filled with black powder so that it was not possible to increase the powder charge. The same held true when cartridge revolvers came into being. The cartridge cases were filled with black powder and the lead bullet rammed down on top. Therefore, if more power were needed, the shooter went to a larger caliber.

Because velocities remained relatively low, shooters did not have too many problems with leading. There was no reason the individual could not cast either soft or hard lead bullets; he could control the hardness of his bullet lead within a reasonable range. Obviously, the softer the lead, the more it will mushroom when the bullet strikes the target. Unfortunately, the softer the lead, the more chance there is of barrel leading. There were and are times when it is advisable to have maximum bullet expansion for a shot or two, thus justifying the use of soft lead bullets.

Lead bullets of varying hardness continue to be used for two basic purposes: hunting big game and target shooting. A big, hard lead bullet is generally superior to a jacketed bullet for big game hunting because it will drive deeply into the animal even though velocity is considerably less than that of a jacketed bullet at the same range.

Target shooting with its accompanying practice demands that the shooter fire thousands of rounds. It would be excessively expensive, without any added benefits, if jacketed high-velocity bullets were used. Lead bullets are readily available and may be cast by any competent handloader for little more than his time and effort. As a result, these bullets are used to provide inexpensive ammunition for practice and sometimes for matches. Velocities are low and barrel lift prolonged. When used within these limits, lead bullets are difficult to beat.

Jacketed bullets came into their own as the result of Jim Harvey's efforts during the early 1950s. They opened up an entirely new field for handgunners. The concept was not new because jacketed rifle bullets had been used since the late 1890s, but the application of this concept to handgun bullets was new. Jim Harvey merely took pure lead, swaged it into a core, dropped it into a bullet jacket and then swaged both core and jacket into a single handgun bullet, thus permitting the use of pure lead (with its ability to expand) plus the high-velocity potential of a jacketed bullet. Put the two together and we have today's high-velocity, jacketed handgun bullets.

During the early stages of development of the jacketed bullet, it was found that if a "half-jacket," or a "short-jacket," was used leading could be kept to a minimum even though the bullet core was pure lead. The forward part of the bullet, being lead, would touch the barrel rifling and lead would commence to coat the barrel; however, the jacket would then swage this lead into a very thin film or even push the lead forward out of the barrel, thus cleaning the barrel. If velocities were held to 1,000 through 1,200 fps and if the barrel were smooth, leading usually presented no problems. As a result, these relatively inexpensive half-jacketed bullets were widely used, and they were also easy to swage and make on many of the better and larger presses. Home swaged bullets became a popular pastime for the shooter.

Unfortunately, like so many good things, these half-jacketed bullets had drawbacks. When the cartridge fired, the hot gases hit the base of the bullet projected by the jacket. The soft lead forward half of the bullet instantly mushed or expanded up to chamber diameter. Remember that the rear of the bullet was already moving forward. When built-up pressures reached a certain level, the front expanded lead portion then slowly moved forward through the front of the chamber. When it hit the forcing cone, far too often the slightly expanded lead nose expanded even more and tried to jam in the tapered forcing cone. The bullet base, held and protected by the jacket, continued to move forward under ever-increasing pressure from the hot gases. At this point, there was no release for the expanding gases ramming against the bullet base, and there would not be any release until the bullet cleared the gap between the rear of the barrel and the front of the cylinder. Occasionally a bullet would stick and jam within the forcing cone, with dramatic results. Sometimes the forcing cone and the threaded portion of the rear of the barrel and the frame would split. Sometimes the top strap and the top chamber would let go. Much depended upon the pressures involved.

When the bullet passed through the forcing cone and down the barrel and out the handgun muzzle there were no problems. Often the gilding metal jacket would stick in the barrel and the lead core would ram out all by itself. This usually happened when too low a powder charge was used, because the friction of the gilding metal jacket was greater than that of the lead core. If too low a powder charge was used with resulting low velocities, these jackets would stick within the barrel with great regularity. Obviously, if the shooter did not know that the jacket remained within the barrel, then the next bullet would produce a lovely big bulge in the handgun barrel. As may be seen, too high a velocity could create problems and too low a velocity most certainly would create problems.

Exterior ballistics had problems also. Remember the tremendous torque of ramming a projectile down a rifled barrel at 1,000 fps. Quite often, the jacket would be revolving at one RPM and the lead core at another RPM. When the bullet finally emerged from the barrel, the jacket would come off the bullet and fall to the ground while the lead core continued on its way. H. P. White, during initial testing for Jim Harvey, ran into this situation rather frequently; reports state that the floor of the range was littered with jackets and that accuracy was very poor.

These run-of-the-mill bullets are still found in quantity. As may be seen, they can be dangerous especially if homemade. If a badly deformed lead slug is found without rifling marks, it is quite possible that one of these bullets was used. Look for the jacket, which should be lying on the ground not too far ahead of the handgun muzzle.

In order to correct both excessive pressures that these bullets could create and to lock on the half jackets, Jim Harvey designed a jacket crimping die. The die could be set to force the forward end of the jacket deeply into the lead core. This would reduce bullet-bearing surfaces to nearly one-half of the normal bullet-bearing surface. Second, crimping the jacket onto the lead core eliminated any chance of the jacket coming off the bullet during passage down the handgun barrel or during the bullet's flight through the air.

At high velocities, these bullets were like explosive bullets because the full potential of the pure lead core, exposed and unsupported at the front half, was completely utilized upon striking the target. The results on game were devastating. Another problem immediately arose. Few handgunners could place their shots into the head or neck of big game. If these bullets struck the shoulder or body, they sometimes expanded so rapidly that there was no penetration. Results varied so much that Jim Harvey decided to go to a long-jacketed bullet, a prototype of what is in common use today.

There were two types of long-jacketed bullets. The first that Jim designed had the conventional bullet shape with the sharp shoulder and recessed base for the bullet nose. The jacket, rather than stopping half way up the bullet body, extended all the way to the sharp shoulder and was then crimped on over the front of the shoulder, thus not only locking on the jacket but eliminating any lead-to-barrel contact. This meant that these bullets could be rammed out at maximum velocities without any barrel leading. The exposed large, pure lead nose would immediately mushroom upon impact. This bullet proved very successful, and huge quantities were sold.

The next step was to produce a bullet that had no sharp shoulder (a holdover from the match target type of bullet) and a really long jacket that ran over the bullet ogive. By running the jacket well past the bullet ogive, there was no need to crimp on the jacket, thus reducing a step in the manufacture of the bullet with resulting lower costs. Once again the bullet nose remained exposed in varying degrees. This was another good seller. Today's bullets are merely minor variations of this original concept.

Almost at once, a new problem arose with these long-jacketed bullets. Handgunners had for over one hundred years used one bullet and varying powder charges to produce the desired velocity and handling characteristics. Therefore, many figured they could fire these jacketed bullets at both high and low velocities. Problems immediately arose. Jacketed bullets are only accurate at high velocities. It took many years before this new type of bullet was generally accepted even though veteran handgunners had proven that two bullets existed and were needed — the low-velocity lead bullet and the high-velocity, accurate, long-jacketed bullet.

Other problems that arose with the high-velocity long-jacketed bullets were parts wear in a revolver and the failure of the cylinder to correctly align each chamber with the barrel throat. For quite a while, this new high-velocity ammunition was condemned as being dangerous and having a tendency to blow up revolvers. It is true that some of this ammunition, both factory and handloaded, was used in older revolvers and revolvers that were not of first quality. Many of these revolvers could not withstand the new, high pressures. Most problems, however, arose from failure of the revolver to correctly align each chamber with the barrel throat. When firing a conventional 158 grain, lead bullet at 775 fps, chamber alignment is secondary. The slow-moving bullet and the low-pressure loading could not create too many dangerous problems. The new high-velocity ammunition left very little leeway for misalignment. As a result, handguns were being blown up — not because the ammunition was dangerous but because the handguns were out of line. With the ever-increasing use of the new ammunition, more and more revolvers were sent back to the factories. I have personally watched nearly one-third of a sixty man firing line fail to complete a course of fire because of improperly cared-for revolvers. Today, with universal acceptance of the long-jacketed, high-velocity ammunition, factories are spending a lot more time checking out cylinder alignment and making certain that it holds.

I have gone into considerable detail because, over the years, many people have come to me with a mashed up hunk of lead in the hopes that I can help them identify it. There are many and will continue to be many half-jacketed and short-jacketed bullets around; it is time that the new generation understand their potentials and their limitations and that such a bullet even exists. Further, in my opinion, it is time that shooters, and especially instructors, understand the basics of lead bullets and of jacketed bullets and the characteristics and potentials of both types.

AMMUNITION USED IN TESTS

Before we commence any firing tests, I would like to specify and list the ammunition that will be used throughout these tests. I have purposely avoided using handloads. If results occur that are unusual, the reader can not blame my handloads and not profit by the results. By using only factory ammunition, I avoid ammunition difficulties, faults, or exceptional results. By using only Remington and Federal ammunition, no reader can accuse me of partiality or of using second grade ammunition.

I do not run these tests with all commercially available ammunition. If the number of rounds that will be fired with one single brand of ammunition in all the various bullet types and loadings were multiplied by all of the ammunition brands obtainable in this country, the project would stall right there. It would be impractical for an individual like myself to attempt this.

If I were to bring in outside help, the chances are we would become embroiled in details and in whether to do this or that.

By doing the entire testing myself, I can delve into minor side issues, try tests that ordinarily would not be run, and most important of all, I can assume all responsibility for both the good and the bad.

I wish to thank both Remington and Federal for their help and for their ammunition. Both have come up with suggestions that will be incorporated in this book.

The following ammunition was furnished by Remington:

Caliber .357 Magnum:
- 125 grain Semi-Jacketed Hollow-Point #R357M1
- 158 grain Semi-Jacketed Hollow-Point #R357M2
- 158 grain Soft-Point #R357M3
- 158 grain Metal-Point #R357M4
- 158 grain Lead Semi-Wadcutter #R357M5

Caliber .45 ACP:
- 185 grain Wadcutter #R45AP1
- 185 grain Jacketed Hollow-Point #R45AP2
- 230 grain Metal-Case — Target #R45AP3
- 230 grain Metal-Case #R45AP4

Caliber 9 mm. Luger:
- 115 grain Jacketed Hollow-Point #R9MM1
- 124 grain Metal-Case #R9MM2

The following ammunition was furnished by Federal:

Caliber .357 Magnum:
- 158 grain Soft-Point Jacketed #357A
- 125 grain Jacketed Hollow-Point #357B
- 158 grain Semi-Wadcutter, lead bullet #357C
- 110 grain Jacketed Hollow-Point #357D

Caliber .45 ACP:
- 230 grain Metal-Case #45A
- 185 grain Metal-Case Wadcutter #45B
- 185 grain Jacketed Hollow-Point #45C

Caliber 9 mm. Luger:
- 123 grain Metal-Case #9AP
- 115 grain Jacketed Hollow-Point #9BP

From time to time throughout this book there will be tests, probably short ones, in which I use only Federal or only Remington ammunition. These tests would provide no additional information by firing both makes of ammunition and would definitely increase the time devoted to firing. Because of this, I do not wish the reader to believe that I am favoring one make of ammunition over the other, or that I am attempting to short-cut tests. This is definitely not the case.

LOAD DEVELOPMENT

The relationship of ballistics to load development work is obvious. Without ballistics, we could not have load development and without load development, accuracy and reasonable pressures could not be obtained. Load development is the assembly of various components and the resulting testing of this specific combination for safe pressures, reasonable handling and accuracy. If these three factors are not present, then a load is basically worthless and possibly dangerous.

As we have seen, cartridges are made up of various components — the cartridge case, the primer, the powder, and the bullet. Each of these components can be vitally important in affecting what a cartridge will do. Few shooters appreciate the time and effort put into load development work by the factories. Actually, it is surprising that ammunition can be as consistently safe and accurate as it is. I would like to discuss this from a very simple approach — that of the average handloader. The same principles apply to what the factories do so that, if the basics are understood, it becomes far easier to understand and possibly correct problems and to also evaluate ammunition when testing it for quantity purchases.

Let's start with an average handgun reloader and the .357 Magnum cartridge. This shooter wishes to have cartridges he can shoot in a match, fire for practice, and use for big game hunting. No single cartridge can correctly accomplish all these goals, so our shooter must sit down with a reloading manual and figure out what he wants to do first and then what components he will use as a starter. When we speak of matches today, we have many and varied types from the conventional NRA matches to the Mexican Silhouette to Doctor Burgess' Laramie match. Being an extremely versatile cartridge, the .357 Magnum can be used for all these sporting events. Let's also assume that our shooter has a Dan Wesson revolver with heavy ventilated rib barrels of four-, six-, eight-, and ten-inch lengths.

With this background, let's start with match ammunition, first for NRA matches using the six-inch barrel that provides weight, stability, and long sight radius.

In NRA matches, we have two basic distances — twenty-five yards and fifty yards for outdoor, conventional, handgun matches. At twenty-five yards, five shots are fired within twenty seconds, and in another match, five shots are fired within ten seconds. This calls for instantaneous recovery after firing, low recoil, superb accuracy. An ideal bullet would be the Lyman lead bullet case with Alloy #2, weighing 115 grains, and indexed as Bullet #358345. This lightweight bullet holds recoil to a minimum and with proper velocity and barrel twist should provide excellent accuracy at twenty-five yards. It may be given a relatively high velocity in the 1100 fps to 1200 fps area without affecting recoil to any great degree.

Our shooter has these bullets checked, weighed, lubricated. Federal once-fired cases — sized, belled, and cleaned in sufficient quantities — are also on hand.

The next problem is what primer to use. Usually if one uses Federal cases, it could be advisable to use Federal primers. The cases are then primed.

The powder and the amount of powder to be used is vitally important. A clean-burning, relatively fast-burning powder would be a good choice. Powder controls the bullet velocity that should be held to a minimum to avoid any chance of barrel leading. Bullseye is a bit too fast burning for such a large capacity case. Generally speaking, it is desirable to fill up as much space within the cartridge case as possible and yet remain within pressure and velocity brackets. For this reason, our shooter selects Unique powder, choosing to start with 6.0 grains behind the little 115 grain lead bullet.

An important and often overlooked detail in match ammunition is the bullet crimp. This final step locks the bullet firmly into the case mouth. True, the case has previously been sized down so that it will, although a tight fit, hold the lead bullet firmly; however, it is imperative that the bullet remain securely held within the case mouth throughout four cartridge firings. When it is struck by the hot gases, it should resist moving with the same pressure as all of the previously fired bullets. Furthermore, being a lead bullet, it should not be deformed in any way when, through build up of pressure on its base, it finally breaks loose from the crimp and starts moving forward. There are two basic types of crimps — the turned-in crimp and the swaged crimp. The turned-in crimp actually forces the front edges of the cartridge case inward to bite into the bullet lead. When the bullet commences to move, the bullet lead must force this turned-in brass outward to permit the bullet to go on its way. There is a tendency for the metal of the case mouth to rip and tear the bullet lead and, in some instances, actually deform part of the bullet as it breaks loose from the crimp. This is not desirable.

The swaged crimp does not force the front edges of the cartridge case mouth inward. It actually sizes down the front two- or three-hundredths of the cartridge case mouth so that this "shrunken" area around the front end of the cartridge case grasps the bullet firmly due to the smaller diameter. When the bullet commences to move, it merely shoves the end of the cartridge case back out to normal size. There is no tearing or ripping of the bullet lead. It is also possible to obtain a much more uniform bullet "pull" with a swaged crimp. Today, dies may be ordered with either standard or swaged crimping shoulders.

Now, let's assume that our shooter has loaded up ten cartridges, observing each step and rule as explained above. Each cartridge contains identical components and 6.0 grains of Unique powder. The handgun is set into a machine rest and the ten shots fired at a target twenty-five yards away. Note that the ammunition is tested in the same revolver it will be fired in during matches — the same revolver and the same barrel.

Group size for the ten shots is perhaps 3 1/8 inches, much too large to be acceptable. Our shooter now loads ten more cartridges using 6.5 grains of powder; abruptly the group size dwindles to 2 1/2 inches. Another ten cartridges, this time using 6.8 grains of Unique; group size shrinks even more to 2 inches. Another ten cartridges with 7.0 grains of Unique are used, but group size is back to 2 3/8 inches. Our shooter loads ten more cartridges with 6.8 grains; again, a tight group of about 2 inches.

At this point, it would appear that this is about the right powder charge, but let's replace the Federal primers with Winchester primers. The groups become larger. Try Remington primers and the group runs slightly under 2 inches. It would appear that our shooter now has his twenty-five yard load. The next thing to do is actually fire it both timed and rapid fire to determine if the recoil and handling characteristics are acceptable. If so, he is set.

His next step is to develop another load for fifty yard slow fire. Doubling the distance and eliminating time — for all practical purposes — permits an entirely different approach. He needs a heavy bullet at moderate velocities in the 750 to 850 fps range. Usually, a 148 to 160 grain wadcutter, lead, sized, and lubricated bullet will prove most satisfactory. In this case the Lyman 358432 cast of Alloy #2 and weighing either 148 or 160 grains could be the answer. Another excellent bullet is the Lyman 358477 that comes in 150 grains, also with a sharp shoulder.

Because recoil is a secondary factor and accuracy a must, and also because a heavy bullet is being used, it is desirable to fill the cartridge case as full as possible with powder. It has been found that a nearly full cartridge case will provide slightly better accuracy than a partially filled case. There are two excellent powders, #2400 and #630. Our shooter should try both powders. He should also work with both regular primers and Magnum primers. Load development routine is similar to the one for the twenty-five yard loading complete with trying, testing, firing, and changing a single component. Finally, some specific combination will be found to do the job well.

Anyone wishing detailed information on lead bullet loadings should read the *Lyman Cast Bullet Handbook.*

While I am discussing lead bullets, I would like to move into the "practice and plinking" areas of handgunning. These are noncritical phases of handgun shooting that are more disciplinary than anything else. They permit the shooter to learn many practical things about his handgun, its handling, and his ability to hit targets. Reject bullets and old primers are ideal for this type of shooting. Everyone has rejects, old components, cases that are ready to split and such things and they may be used with light loadings for shooting at cans, practicing certain techniques, and for honing reflexes while trying out new concepts. Using .357 Magnum cases, 3.5 to 4.0 grains of Bullseye makes an excellent powder charge for use with many weights of lead bullets.

Now I want to get into the high-velocity, jacketed bullet phases of handgunning. Today we have several courses of fire that require high velocity combined with extremely good accuracy. This bears no relation to the lead bullet loadings discussed previously. It is a whole new field of load development. I refer the reader to the new Sierra reloading manual.

The Mexican Silhouette, to pick a single modern match, demands many things from the shooter. He must first have a handgun that is capable of producing the finest accuracy. It must further be able to withstand near maximum pressures for long periods of firing. At this time, the 170 grain, .357 Full Metal Jacket Silhouette bullet is recognized as *the* .357 bullet for the Mex. All other .357 Magnum bullets lack the inherent potential of this specific bullet to knock over the steel targets. Therefore, I want to concentrate upon this one bullet. We now enter a field that requires slow-burning powders, the use of the ten-inch Dan Wesson barrel, and maximum accuracy. Time is of no importance so recoil is not a critical factor.

It is advisable to use only cases that are in perfect condition. Only Magnum primers need be considered. Our shooter has the Sierra 170 grain Silhouette bullet so all he needs at this point is to select a powder and commence testing. While this may appear simple, it is not quite that simple because he will be working with a ten-inch barrel allowing him to use extremely slow-burning powders. On the other hand, a too-slow-burning powder may not prove satisfactory. The ideal is to develop a loading that will burn all the powder within nine inches of barrel length. This would utilize the full potential built into the powder charge. If the powder burns in seven inches, our shooter would be wasting energy because the bullet would have no pressure behind it after moving seven inches down the barrel. The last three inches of barrel length would only serve to reduce velocity through friction. Conversely, if the powder charge cannot burn within a ten-inch barrel, then our shooter is blowing unburned powder out of the handgun muzzle and wasting energy.

The Mex demands velocity. Without velocity the bullet cannot knock over the steel targets. As a result, it is necessary to spend time determining just how efficient a powder charge really is and this can be done by evaluating velocities and the amount of unburned powder that leaves the handgun muzzle. There are basically about four powders suitable for use with this 170 grain Silhouette bullet. They are Blue Dot, #630, H-110, and #296. Near maximum loadings are necessary; there is nothing to be gained by commencing powder charges far down the line. If maximum loads are 15.0 grains, then start loading around 13.5 grains. Somewhere in that top bracket a satisfactory loading will be found that will still give acceptable accuracy.

When loading for the Mex be certain that a conventional crimp is used to lock the bullet solidly into the cartridge case. A cannelure is provided. If the bullet pulls or moves during the recoil of firing other cartridges, problems will arise and accuracy can suffer. In my opinion, an efficient loading can be determined by watching velocity figures carefully. As long as bullet velocity shows no indications of dropping, the shooter is on the right track. By that I mean that if velocities with ten-inch barrel move steadily from 1100 fps to 1135 fps to 1160 fps to 1162 fps and then appear to level off, there is a good indication that maximum velocities have been reached.

Once this has happened, leave the handgun locked in the machine rest and spread newspapers out in front of the handgun muzzle for a distance of at least ten feet. Then fire ten or twelve or more shots. Without disturbing the paper, examine the surface looking for particles of unburned powder. If you find any amount of unburned powder, you will know that the powder is too slow burning for that specific barrel length — providing, of course, that your loadings are close to maximum. Drop back to a slightly faster burning powder and repeat the load development work moving up in half grain powder charge increases. When you reach a full grain below recommended maximum powder charges, cut your powder charge increases down to one quarter grain and move up very slowly from there.

Thus, by carefully watching both velocities and the amount of unburned powder leaving the handgun muzzle, it is possible to arrive at what should be a practical and efficient loading for the Mex. Beware of exceeding maximum recommended loads. All top quality handguns have a considerable amount of safety margins built into them but constant and repeated firings with over maximum loadings can cause many problems. Frames have been bent, cylinders wrenched out of line, and handguns eventually blow up. There is no reason to abuse good equipment. If you cannot accomplish your goal with your .357 Magnum, then go to .41 Magnum or .44 Magnum.

Because accuracy remains a dominant factor in the Mex, some give and take will be necessary between accuracy and velocity. As one shooter remarked, "if you cannot hit 'em, you sure as hell are not going to knock them over." Therefore, be willing to sacrifice a bit of velocity to pick up a bit of accuracy, within reason of course.

So far, we have discussed matches where accuracy is absolutely necessary; by accuracy I refer to standards that would have been considered impossible to attain twenty years ago. Now, let's move into another phase of handgunning where accuracy is secondary to knock-down power and bullet performance at the target. In match shooting, bullet performance is not a factor. The moment a lead bullet passes through a target and makes a nice clean hole in the desired spot, its usefulness is past. Similarly, when the 170 grain full-metal-jacket Silhouette bullet strikes the metal target and knocks it over, its usefulness is past. Not so in the hunting fields. An entirely new dimension has been added to load development work. It is known as "stopping power — knock down power — bullet performance." In the hunting fields, the handgun/bullet combination must be capable of placing the bullet in the desired spot on the animal so that reasonable accuracy is necessary. The moment the bullet impacts, it must do its job: expand, drive down into a vital spot, and impart all of its inherent impact shock into the animal. Our 170 grain Silhouette bullet is useless for hunting. So are most match lead bullets. It is necessary to have a high-velocity, jacketed expanding bullet of around 150 to 160 grains in weight.

The Dan Wesson revolver with an eight-inch barrel is ideal for big game hunting. When working up a load for use with this handgun, all of the regular load development work must be carried out, plus making certain that the powder is being burned within the eight inch barrel, plus determining what the bullet will be expected to do at hunting ranges. Bullet performance is the additional factor that must be carefully evaluated prior to taking the handgun/ammunition combination into the hunting fields.

I have discussed this at some length in other places in this book, so I will not take the time and space to repeat myself here. I do wish to impress upon the reader that load development is an important part of ammunition production, whether done by the factory or by the individual reloader. It is possible, given unchanging and specific requirements, to produce a loading that will handle nearly any specific situation or any specific target. Change one detail and the ammunition potential may not do the job. This is a difficulty encountered in ammunition production. As an example, let's assume that all men are five foot, ten inches in height and weight 170 pounds, walk around in bathing trunks and are constantly at all.times in a gentle, happy mental state. Under these conditions a .22 caliber long rifle cartridge fired in a six-inch barrel pistol would dispose of them without difficulty. But change one detail such as the mental state or the size or the clothing and we could need .44 Magnums to dispose of these people.

To carry this one step further, assume that we have a partition. We have developed a load that will constantly and efficiently ram a bullet through this partition. We are quite happy about all this. Let someone place a sheet of 16 gauge steel on either side of this partition and we discover that the bullet can no longer produce the desired results. This is why we have load development. It permits the development of a loading that will do the job under basic conditions and at basic distances. Just as a match target twenty-five yard loading is useless for firing at 200 meters in the Mex match, we must be able to evaluate a firing situation and then select the correct ammunition/handgun combination to do the job efficiently.

Far too many people believe that because a certain loading in a certain handgun will produce desirable results at ten or twenty feet that this same loading will prove ideal when firing at a target that is now out at fifty-seven yards. Velocities fall off the moment the pressures cease to force the bullet to go faster, whether this is within the handgun barrel or when the bullet leaves the handgun muzzle. A loading that works ideally at twenty feet can be ineffective at forty-one yards.

The choice of bullet and bullet velocity controls what happens at the target. A high-velocity, lightweight bullet of perhaps 100 grains can go to pieces upon impact with the target, thus producing little, if any, penetration at the target. Another heavy bullet at lower velocities can slam through a target and continue on indefinitely without imparting more than a fraction of its impact shock to the target. Still another bullet can strike the target and because of its design drive down deeply within the target and expand and impart all of its potential impact shock to the target. Thus, it is impossible to produce a single loading that will do all things. It is, however, possible through testing and evaluation and load development to come up with a handgun/ammunition combination that can probably do the job better than any other loading.

I have delved into details and the load development picture so that the reader may better understand basic ballistics. The two are tied together. It is impractical to have the one without the other. Unfortunately, there is no simple solution to ammunition problems and deficiencies. Everything must be a compromise. It remains up to the individual to select the compromise that as a result of exhaustive testing promises to do the job better than any other.

Chapter 2

REVOLVER AND PISTOL MECHANISMS

THE REVOLVER

REVOLVERS are quite complicated mechanisms and, from a practical point of view, entirely obsolete in today's world. A revolver consists of a frame of some kind within which is suspended a cylinder. This cylinder contains five or six chambers. The cylinder revolves, actuated by a series of internal parts, usually under the control of a trigger that moves a hand that revolves the cylinder a predetermined distance. When the trigger is brought back a specific distance, the cylinder is then supposedly locked into position with the top chamber lined up with a barrel that is somehow fitted into the frame directly ahead of the top chamber of the cylinder. Added pressure to the trigger usually trips a sear, and this allows the hammer to fall, thus firing the cartridge.

Due to manufacturing tolerances, wear, use, and other details that are impractical to control beyond normal quality controls, revolvers will usually have something slightly out of line, loose, etc. Consider how difficult it must be to perfectly line up six holes with a single barrel throat every time for one hundred thousand shots, and then multiply this by twenty million revolvers, and it is possible to appreciate the problems inherent in revolver manufacture. Revolvers are a compromise, and it is amazing how well they perform and how very well many of them shoot.

When the manufacturer drills six thousand holes per day in revolver cylinders, and then attempts to fit one thousand cylinders so that they line up with one thousand barrel throats, we run into tolerances. These tolerances must exist, or the cost of producing revolvers would increase tenfold.

First, the cartridge must fit into the cylinder chamber, not just one type of cartridge, but every cartridge ever made and every cartridge that will ever be made for that specific caliber.

Keep in mind that ammunition manufacturers also have their practical working tolerances. Their problem is to maintain outside measurements *under* predetermined specifications. An oversize cartridge cannot generally be chambered. Conversely, revolver manufacturers must keep their chambers slightly oversize, rather than undersize, so that they can accommodate the occasional oversize or full size cartridge.

When the cartridge is chambered, the revolver action locked, and the trigger pulled, the firing pin falls and strikes the primer. This blow upon the rear of the cartridge case tends to force the entire cartridge forward within the chamber. Remember that this chamber may be slightly long, thus allowing a certain amount of play. Also bear in mind that, while the revolver cartridge is a rim cartridge, the distance between the rear of the cartridge head and the front of the recoil plate in the frame can vary, again due to tolerances. True, the cartridge itself cannot go any further into the chamber than the rim will permit, but it may have been seated all the way back when struck by the firing pin. This slack will be taken up somehow during firing.

In order to chamber the cartridge, the chamber that actually holds the cartridge is larger than the bullet diameter. Ahead of this portion of the chamber is a smaller diameter area through which the bullet will pass; this area is usually larger than the bullet diameter, again, for reasons of production tolerances. The average revolver has six of these chambers in each cylinder.

Once the powder charge ignites, the cartridge swells under the ever-increasing pressure, then it locks into place. The cartridge case itself merely holds the components and has never been designed to contain the internal gas pressures by itself. This is done by allowing the case to expand until it meets the walls of the chamber that, in their turn, control and hold the gas pressures. At the rear, the cartridge case is supposedly held by the recoil plate through which the firing pin protrudes and which remains embedded in the rear of the primer.

In the front end of the cartridge case is the bullet, which is securely held by the case mouth. When the case expands and the pressures commence to bear upon the bullet base, the rear of the bullet tries to expand due to the inertia of the projectile and the force required to break the seal at the case mouth and start the bullet on its way. The bullet base cannot expand any further than chamber dimensions permit. It then starts forward. Depending upon the quality control governing the reaming and polishing of that specific chamber, the bullet may continue to expand slightly or remain the same or even be swaged down a bit by the tapered cone forward of the actual cartridge chamber.

In order to prevent the seating of .357 Magnum cartridges within a .38 Special chamber, the cartridge chamber is reamed out only enough to accept an average .38 Special cartridge. Ahead of this cartridge chamber is the balance of the cylinder chamber that is supposedly bullet diameter and through which the bullet will pass to reach the forward end of the cylinder.

Whatever the bore size, the bullet will expand to this dimension, with the base expanding and moving faster than the nose. The bullet moves through the area forward of the cartridge case, expands fully, and reaches the forward end of the chamber. This is one of the crucial areas. Remember that the cylinder turns and, each time it turns, supposedly lines up one chamber with the barrel throat. The rear of the barrel may or may not be tapered. This short cone-shaped area is free of rifling and should be highly polished. It is called the forcing cone. This — in theory — is perfectly aligned with the front of the chamber. Between the rear of the barrel and the front of the chamber, there must be a gap. Without the gap, residue, lead and fouling would build up and prevent the cylinder from turning freely. Even with the gap, many times the front of the cylinder will bind against the rear of the barrel.

The bullet nose emerges from the front of the cylinder chamber. It is still trying to expand under the pressures that continue to ram the base forward. In the fleeting instant that the bullet nose is free in the air within the gap between the barrel and the cylinder, it tries to expand but slams into the forcing cone somewhere within the forcing cone. The balance of the bullet moves through the gap (also attempting to expand) and is rammed into the forcing cone, caught by the forcing cone, and swaged down by the forcing cone. If the forcing cone is not true with the front of the chamber, particles of lead and residue will be flung out the sides of the gap and also backwards by the barrel face that surrounds the rear of the forcing cone. This back splatter is well known to revolver shooters.

If the forcing cone is not true with the chamber, then the bullet will be forced to accommodate to the angle of entry. By this time, the bullet nose has hit the rifling in the barrel and is trying to turn, even though the rest of the bullet continues to come through the forcing cone. Quite often, the bullet will hit the rifling and actually skid for an inch or two within the rifled barrel before the rifling can grab the bullet and force it to rotate. This can be easily verified by firing high velocity bullets through a two-inch, or one-inch, barrel into cotton waste. Lead bullets will skid because modern day rifling is not cut to handle lead bullets — rather, it is a compromise that can handle both lead and jacketed bullets. Usually modern rifling is both too wide and too shallow to correctly grab lead bullets.

By this time, the bullet, regardless of its size when it hit the forcing cone, is now swaged down to barrel size and should be riding the rifling and filling the barrel grooves. Few shooters appreciate the tremendous torque that the bullet is subjected to, or to the speed of rotation. Many years ago, when I was in the bullet and ammunition business, I discovered that it was necessary to use a relatively thick jacket or the rifling plus the torque would rip the jacket to pieces.

If all has gone well so far, the bullet base has retained its shape to present a constant base to whatever gases remain — these gases exert pressure against a uniform surface. The balance of the bullet should be swaged to a constant size without having been deformed. Any deformation that might have occurred up to this point should be ironed out. Further, the original center line of the bullet should be true with the center line of the barrel. This is the straight line that runs through the center of the bullet nose, down through the center of the bullet and out through the center of the base. Any off-center bullets will lack the ability to provide consistent accuracy.

If the correct powder has been used, pressure will remain upon the bullet base as it leaves the muzzle of the revolver. This is very critical because the escape of gas along one side of the bullet base prior to the bullet actually leaving the muzzle can tilt the bullet base, forcing the bullet to fly at an angle and not on a direct line to the target. No gas should escape past the bullet once the bullet is seated in the rifling; not until the bullet base leaves the muzzle should gas appear. As the bullet leaves the muzzle with full pressure behind it, then, and only then, should gas leave the handgun muzzle. A perfect fit of bullet to barrel to rifling is necessary to do this.

When we consider all the punishment that the bullet takes during this passage from the end of the cartridge case to the muzzle, it is a wonder that we can obtain any reasonable accuracy.

The Dan Wesson Revolver

Before we progress any further into interior ballistics, we should review our working tools. The basic tool that will be used throughout most of the tests is the Dan Wesson revolver, .357 caliber Magnum. Normally, when we wish to test ammunition in a two and one-half, four-, and six-inch barrel, we must have three different handguns, each with the desired barrel length. There is nothing wrong with this, and we will obtain velocity readings and accuracy results consistent with what each handgun will produce. Bear in mind, however, that due to manufacturing tolerances, there will remain definite and individualistic variations between these three revolvers controlled only by the quality of manufacturing controls exercised during the manufacture of these three revolvers. These variations will, to some degree, affect both bullet velocity and bullet accuracy.

The Dan Wesson revolver design is unique in that it permits the easy and simple replacement of the barrel, hitherto a difficult and sometimes potentially dangerous process with most other revolvers. This revolver is the creation of Daniel B. Wesson, a direct descendant of the original Daniel B. Wesson who founded the Smith and Wesson firm in the middle of the nineteenth century. When Bangor Punta took over Smith and Wesson in the 1960s, Dan Wesson was production manager.

Figure 2-1A. The Dan Wesson revolver fitted with the six-inch barrel. The four-inch and two-inch barrels are shown.

Figure 2-1B. The Dan Wesson revolver with the barrel removed.

Because of his personal interest in firearms, plus his knowledge of machine tools, production, and shop routine, Dan knew firearms from the ground up. He had literally grown up in the middle of machine tools. Shortly after the purchase of Smith and Wesson by Bangor Punta, Dan left Smith and Wesson. For several years he toyed with the idea of manufacturing an entirely new type of revolver — one that would be modern, flexible, and simple to produce. The trend from cutting steel to the use of investment castings was strong at that time. Today, it is a widely used process. It permits the use of parts designed to do the job — not designed to be easy to manufacture out of cut and machined steel.

I have used the new Dan Wesson revolver for many years, due to the easy interchangeability of the barrels, and have been able to conduct many formerly impossible tests with it.

We now have a single revolver frame within which is a single revolver cylinder bored with six chambers to hold and fire the .357 Magnum cartridge. This same unit will be used without change or alteration. This provides a constant that is not present when we use three, four, or even five revolvers. This increases the validity of the tests.

When we need a two-, four-, or six-inch barrel, the desired barrel will be screwed into the frame and locked into place.

Perhaps I should explain how this is done. When completely assembled, the Dan Wesson revolver appears quite conventional; this is deceptive. First, the barrel itself is a thin tube that screws into the front of the revolver frame. Over this actual barrel is fitted the shroud that goes on from the front and butts solidly against the front of the frame. This shroud is then locked into place by means of a screw at the bottom of the shroud, and also by means of the lock nut at the muzzle end of the barrel. Thus, neither the barrel or the shroud can move sideways or forwards or backwards. Both are securely locked into place as one unit.

The question will immediately arise as to the gap between the rear of the barrel and the front of the cylinder. When the barrel is screwed into the frame, a shim of the correct width is placed between the rear of the barrel and the front of the cylinder so that when the barrel touches this shim, the distance from the front of the cylinder to the rear of the barrel is correct. The shroud is then slipped over the barrel, the bottom shroud screw locked into place, and the muzzle nut tightened. The shim is then removed from between the cylinder and the barrel and the revolver is ready to be fired.

The entire operation of barrel changing takes two or three minutes. If instructions are followed, it is foolproof. I consider this to be extremely important when attempting to run this type of test.

Part of the Interior Ballistics testing that I will report shortly involves firing the Dan Wesson revolver without any barrel. This is important for two reasons. First, it will show bullet condition during one phase of interior ballistics. Second, it will illustrate what every professional ballistician knows — that men can be killed by revolvers without barrels. Accuracy, obviously, is poor and definitely limited as to distance, but it is a definite and practical solution to disposing of an unwanted person without having to worry about barrel markings. On the other hand, chamber and recoil plate, plus firing pin characteristics, must also be altered if the fired cartridge cases are not immediately destroyed.

The ballistician should always bear in mind that, with such a ready and rapid barrel changing capability, identification marks on bullets may require far more work and problems than with conventional revolvers. However, this is nothing new in the handgun field because we have long had pistols with easily replaceable barrels.

In my opinion, the Dan Wesson revolver is ideal for use for the proposed tests. I can also recommend it highly as both a police and sporting arm.

PISTOLS

Pistols treat bullets far more leniently, for the simple reason that the pistol cartridge is seated within the one chamber in the pistol barrel. There is no lining up of chambers to barrel throats. The cartridge is fed directly out the magazine, forward up a ramp, and then slammed into the barrel chamber and held there by the face of the breech block: simple, yet effective.

When the cartridge fires, the bullet passes directly out of the chamber into the barrel without jump, gaps, or any misalignment. Some inexpensive barrels may have some chambers cut out of line with the center line of the barrel, but normally this is not true. On the other hand, due to manufacturing tolerances, some chambers may be slightly oversize and others very snug, but basically, the pistol bullet has a distinct advantage over the revolver bullet in its lack of abuse and punishment. It goes directly into the barrel, is grabbed by the rifling that has been designed to handle jacketed bullets, and moves without interruption through the barrel and out into the air.

It would be logical to assume that pistols would, therefore, be far more accurate than revolvers. Pistols also have problems that affect accuracy. Neither one, when correctly and carefully assembled, has any great accuracy advantage over the other, although the revolver with its long barrel is far superior for long range match shooting and for hunting.

The Thompson/Center Contender Pistol

The Contender is not a new concept. Single shot pistols have been around for hundreds of years. They are still used in Olympic handgun matches and it was only recently that Webley and Scott, the famous British firearms firm, dropped their single shot match pistol. What is unusual about the Contender is that it is so designed that unlimited numbers of barrels, of many different calibers, may be fitted to a single receiver. Barrel calibers run from .22s through .45s.

The one that will be used for these tests is a .357 Magnum with a ten-inch barrel, which includes the chamber. Because of the design of the Contender, it provides us with a solid breech pistol.

Revolvers have the gap between the cylinder and the barrel; the Contender does not. A revolver has the six chambers in the cylinder; the Contender does not. The cartridge in the Contender lies within a chamber in the barrel as is the case with the

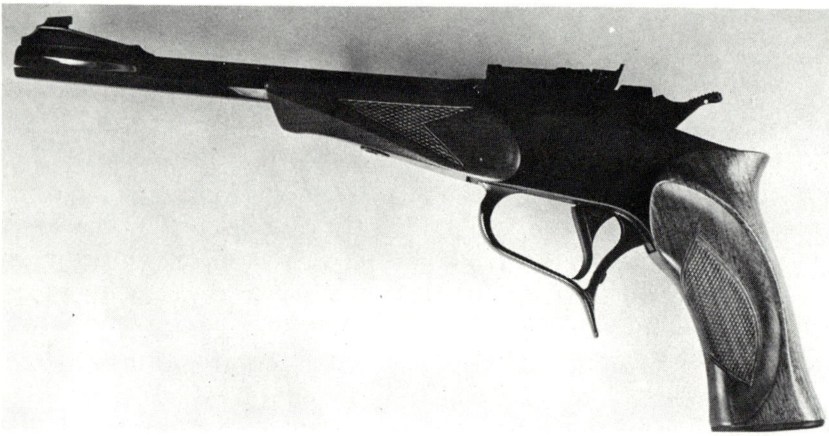

Figure 2-2. The Thompson/Center Contender.

semiautomatic pistol. However, the Contender does not have the complications and potential trouble spots of the pistol so that, in theory at least, the Contender should be ideal for comparing velocities and accuracy out of both the solid breech pistol and the revolver. Thus, the two entirely different mechanisms may be used to evaluate ammunition of the same type, in this case the .357 Magnum.

THE MISUSE AND MISLEADING POTENTIALS OF HANDGUN MECHANISMS

Not too long ago, there was quite a bit of publicity concerning those people who were capable of studying a computer mechanism and its programming and then by careful manipulation force the computer to extract information or credit certain accounts with small but constant amounts of cash. When exposed, the criminals were accused and convicted.

Similar situations but with entirely different results are cropping up in the firearms business, and I believe that a warning is long overdue. Both law enforcement personnel and prosecuting attorneys appear to be failing to correctly evaluate specific situations, thus permitting criminals to go free on technicalities that should have been foreseen. Unfortunately, many people are not aware of some of these details and the potentially misleading situations that can arise. Designers and executives of firearms firms often believe that safeguards built into firearms have been so perfected that they are definitely fail proof. Their minds have been so conditioned over the years that they are incapable of realizing in advance that sometimes, under certain conditions when studied and examined with care by professional gunsmiths and/or people who enjoy examining mechanisms, these mechanisms can fail. Let me discuss a few specific situations.

Few people will challenge the statement that pistols with magazine safeties built into them will not fire if the magazine is removed. This is basically a true statement. Before we get into this, let me remind the reader that most professional agents, plus a great many law enforcement officers, will have their pistols customized by gunsmiths. One of the things that the gunsmith will do is remove the magazine safety from the pistol mechanism. This is done so that the agents can fire the pistol even though the magazine has been dumped and the agent is reaching for a loaded magazine. It removes a moment of pistol inoperability that exists with every pistol fitted with a magazine safety. This brief moment in time when the magazine safety takes over could represent the difference between an agent's survival or an agent's death. For this reason, there are many pistols today that have had the magazine safety removed. This is a legitimate and practical modification of a pistol mechanism. Do not confuse this situation with what I am now about to discuss.

As far as I know from personal examination, every pistol fitted with a magazine safety can, under certain conditions, be fired with the magazine removed. Take a Smith and Wesson Model 39 pistol. Remove the magazine and the pistol cannot be fired. This is absolutely correct. The trigger mechanism has been disconnected. Now replace the magazine in the pistol butt. Cock the hammer. With the trigger finger apply a light, steady pressure upon the trigger. With the left hand, press the magazine release and remove the magazine from the pistol butt, retaining the pressure upon the trigger. With the magazine completely out of the pistol, apply normal pressure to the trigger and the pistol will fire.

When the hammer is cocked, the trigger mechanism prepares to fire. Then, when the slight pressure is applied upon the trigger, the entire trigger mechanism is held in the "ready to fire" position through a small, but more than sufficient, amount of friction. This friction continues and holds the trigger mechanism in the "ready to fire" stage even as the magazine is being removed from the pistol butt. At this point, if pressure is applied to the trigger, the trigger mechanism will respond in the normal manner by firing the pistol. The trigger mechanism has never been rendered inoperable by the magazine safety. However, if, at this point, the pressure upon the trigger is removed ever so lightly, the magazine safety device will instantly take over and make the pistol incapable of firing until the magazine is reinserted into the magazine butt and slammed home.

The ability to handle the preceding magazine safety circumvention takes an intimate knowledge of the mechanism and indicates an in-depth study of the pistol. I would suggest that any law enforcement officer or district attorney who is faced with this as a defense in court (pistol fired with magazine removed) undertake an exhaustive and painstaking investigation of the defendant and the defense attorney.

Over the years many officers have come to me with the complaint that their revolver jams. Double action revolvers in perfect condition are not supposed to jam, and "expert" witnesses and others will testify in court that such cannot happen. Unfortunately, it can! Most revolvers have a point at which it is simple to jam the trigger. Pull back the trigger all the way until the hammer falls. Then slowly permit the trigger to move forward. The amount of forward movement depends upon the design of the revolver mechanism but at some point short of being fully forward, the trigger may be jammed by pulling it to the rear. Anyone who has done much fast shooting with a double action revolver has jammed the trigger This is due to the speed with which the trigger finger moves forward and then attempts to again cycle the revolver mechanism for another shot. If the finger moves faster than the revolver mechanism, the trigger can jam. This jamming can be done deliberately without any problem or difficulty, but like circumventing the magazine safety, it is not a normal nor anticipated part of the cycle of a revolver mechanism. I have known officiers who told me that their revolver "jammed." Others mentioned that they had seen felons attempting to fire but were unable to get off a shot and yet the handgun would be found to be in perfect mechanical condition. This is a situation that could carry potential problems.

Every now and then some person is shot because the "handgunner" applied pressure to the trigger at the same time slowly pressing down on the pistol safety. Obviously when pressure is applied to a pistol trigger, all that stands between the pistol firing and not firing is the manual safety. If this manual safety is being pushed off, at a certain point it will release the hammer or some other part of the trigger mechanism and allow the hammer to fall, thus firing the pistol. It is impractical to build into a handgun fail-safe devices such as we have on fighter pilot seats to prevent the accidental ejection of the pilot from the plane. If a pistol safety is pushed into the OFF position, it would be abnormal if the pistol would not fire when the trigger is pulled, yet situations arise in court with this situation being used as a defense.

I hope that this brief discussion will alert and aid law enforcement and the legal profession. In my opinion, far too many homicide cases go into court without adequate and in-depth preparation when dealing with firearms.

I have dealt with ammunition adapters and with the ability to replace and alter barrels and handgun characteristics. All of these details should be kept in mind when trying a person for homicide.

Today we have so many fine custom barrel makers and so many people adept at using machine tools that it is no longer always possible to examine a recovered bullet and state that it was fired from such and such a handgun. There is no substitute today for long and painstaking investigation when someone's life is at stake.

Similarly, the use of paper-wrapped bullets fired from a cartridge of entirely different caliber is a very simple process. Using a .44 Magnum cartridge, it is possible to wrap a .41 Magnum bullet in paper, insert in into a .44 Magnum cartridge case, crimp the bullet and paper tightly, and then fire it. The bullet could never be traced to the specific handgun that fired it.

The .350 caliber Remington Magnum cartridges fires a bullet the same diameter as the .357/.38. Many owners of the potent .350 Remington Magnum rifle make up varmint-plinking low-velocity loads using cast lead handgun bullets and also jacketed handgun bullets. Such a bullet fired from the rifle can confuse a homicide investigation.

The list of similar modifications is long and other combinations can go on up through the big .45-70 cartridge.

Fortunately, the criminal element is not usually well versed in handloading or in the use of components and firearms. Nevertheless, it is important for law enforcement to evaluate a situation that appears baffling and to discuss it in depth with all its ramifications with knowledgeable professional shooters and handloaders.

Ordinary people who are too knowledgeable regarding mechanisms and ammunition should be suspect, especially if standing trial for homicide that appears to be accidental.

Chapter 3

THE REVOLVER

FIRING WITHOUT A BARREL

I HAVE briefly touched upon bullet base expansion. Bullet base expansion takes place at relatively high velocities. It is possible to carefully measure wadcutter match bullets with solid bases and discover a slight amount of base expansion. Expansion with hollow-base bullets is greater but difficult to see. This is because of low pressures, low velocities, and hard bullets that resist deformation. This very minor amount of bullet deformation may be one of the contributing factors towards producing top accuracy.

I had hoped during these tests to be able to determine bullet velocity. It would be interesting to know how much velocity is gained or lost by a bullet when fired out of a chamber only. Does the barrel really reduce velocity? Unfortunately, it was impossible to get a bullet to register on the second chronograph screen. Much of the time, we could not keep a bullet within the scope of the first Oehler Skyscreen, much less the second which was ten feet further down range.

As I mentioned previously, the hot gases and pressures that result from firing a cartridge bear directly upon the bullet base. The base tends to move forward before the bullet nose starts to move, thus expanding the bullet base up to chamber dimensions. As the bullet leaves the chamber to enter the air, additional expansion can occur, thus forcing the bullet base to grow even more. Many shooters do not understand this. In order to factually show how much expansion there can be, I removed the barrel from the Dan Wesson revolver and fired the bullets into Duxseal from a distance of about three to four feet. The Duxseal was contained within a metal coffee can. The fired bullets were the dug out of the Duxseal and the Duxseal closed up again to form a solid mass.

Figure 3-1. Firing into Duxseal with the Dan Wesson revolver without a barrel.

Figure 3-2. The revolver has just fired.

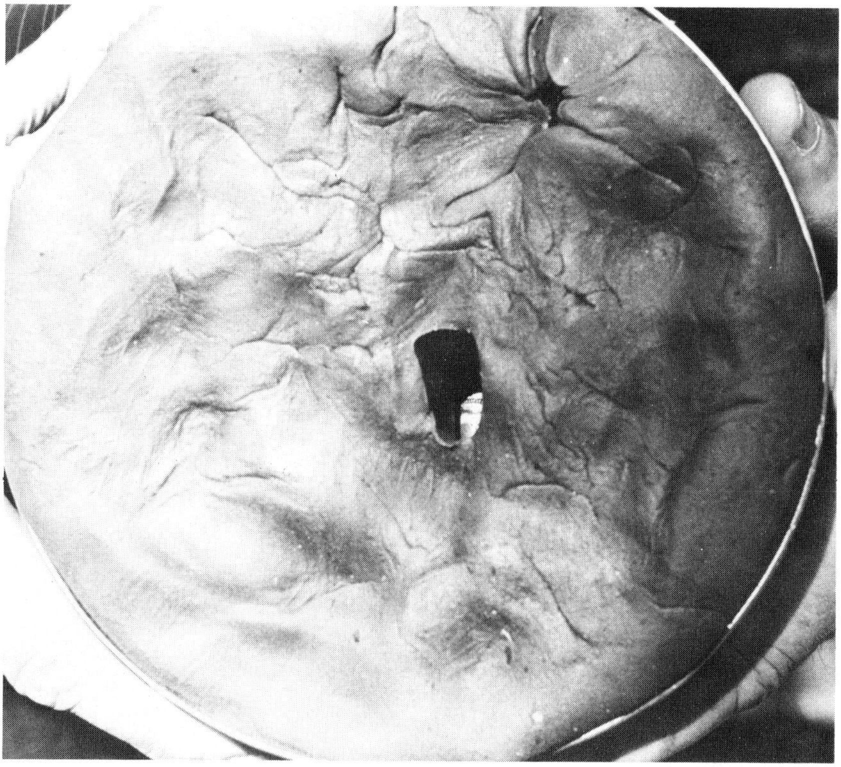

Figure 3-3. Even at this short distance, the bullet turned sideways in the Duxseal.

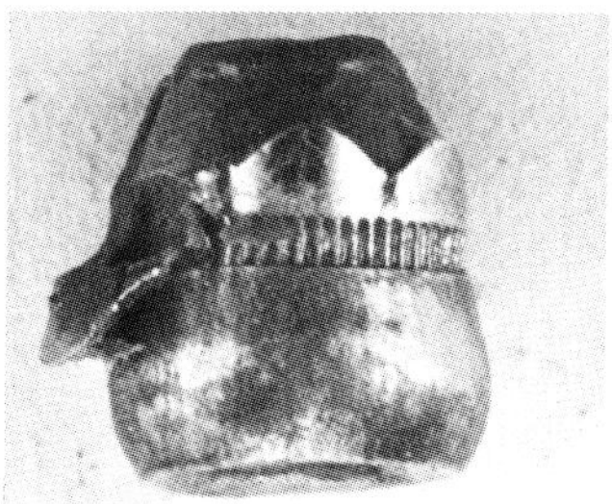

Figure 3-4. Remington .357 Magnum 125 grain, semi-jacketed, hollow-point bullet fired in the Dan Wesson revolver without a barrel.

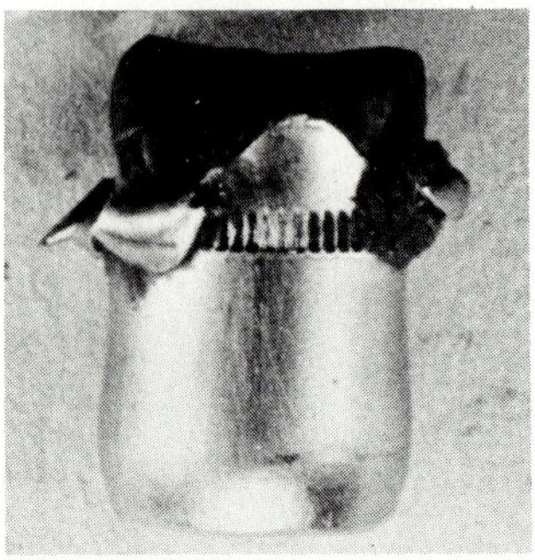

Figure 3-5. Remington .357 Magnum 158 grain, semi-jacketed, hollow-point bullet fired in the Dan Wesson revolver without a barrel.

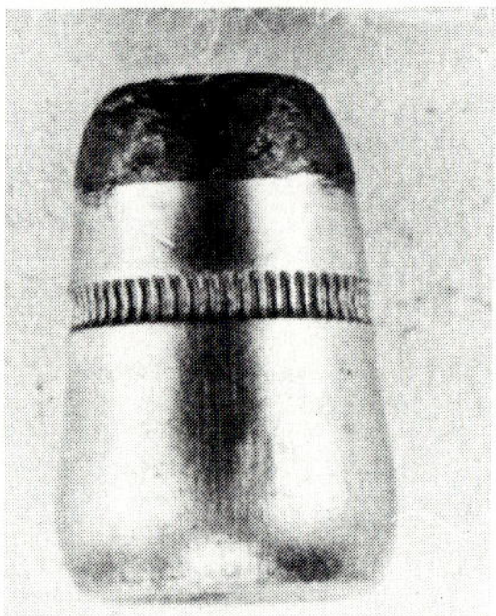

Figure 3-6. Remington .357 Magnum 158 grain, soft-point bullet fired in the Dan Wesson revolver without a barrel.

Figure 3-7. Remington .357 Magnum 158 grain, metal-point bullet fired in the Dan Wesson revolver without a barrel.

Bearing in mind that an unfired .357 Magnum bullet measures approximately 0.357, it is interesting to note that recovered bullets had their bases blown up to as large as 0.691. The results are shown in Figures 3-4 through 3-8.

Figure 3-8. Remington .357 Magnum 158 grain, lead semi-wadcutter bullet fired in the Dan Wesson revolver without a barrel.

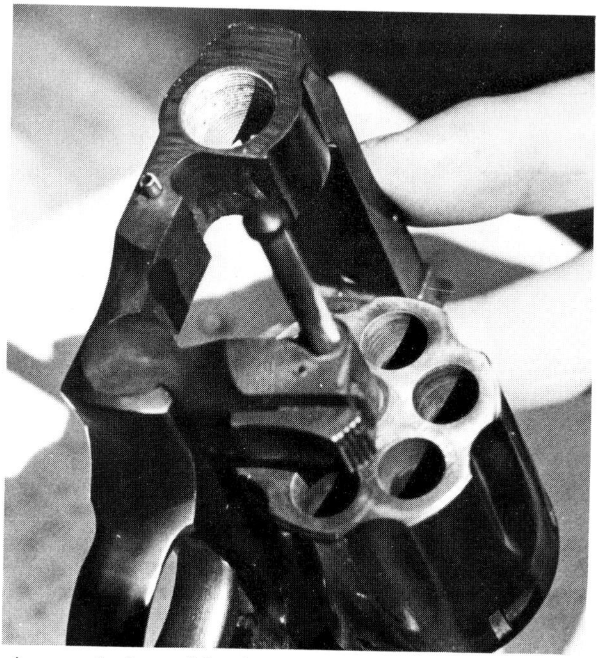

Figure 3-9. Accumulation of lead and fouling in the threads of the frame.

An unanticipated side issue came to light when we examined the barrel threads in the frame of the revolver. The threads had been so filled with lead that the barrel would not screw into the frame. The threads had to be carefully scraped out, wire brushed, and cleaned many times before they would accept the barrel. At first, we thought this was due to the spraying of hot lead, but upon closely examining the fired bullets, we could note that the expanded bullet bases showed definite indications of having literally scraped against the barrel threads in the frame.

This could be of interest to police laboratories when a homicide is thought to have been committed with a revolver without a barrel. If they remove the barrel and discover bits of lead on both the barrel threads and the threads in the frame, the odds are that the revolver was fired without a barrel.

THE BULLET AND THE BARREL

Now that we have seen what happens to a bullet fired in a revolver cylinder, let's examine what takes place when a barrel is screwed into the revolver frame. The rear of a barrel should not touch the front of the cylinder. Sufficient space must be left between the rear end of the barrel and the front of the cylinder so that leading, powder residue, and other surface coatings will not build up and force the cylinder to stop rotating. Smith and Wesson and some government gunsmiths have, at various times, produced revolvers with no gap at all between cylinder and barrel. These revolvers were specially fitted so that silencers could be attached to the end of the barrel; however, these handguns were not expected to fire more than one or two shots. When using relatively low-velocity ammunition (which is necessary when using a silencer) perhaps four or five shots could have been fired before the cylinder locked up from residue and lead build up. Because the gas could not basically escape from the cylinder with its accompanying noise, plus the silencer on the muzzle end of the barrel, these revolvers proved useful. This was an interesting experiment yet one that was eminently impractical. Normally, there must be an obvious gap between the front of the revolver cylinder and the rear of the barrel.

The rear of the barrel is reamed out to provide a funnel into which the bullet nose enters as it leaves the cylinder chamber. This forcing cone, as it is known, can be shallow or deep and it varies with the handgun manufacturer's specifications. There is no rifling in the forcing cone. Barrel rifling commences at the front end of the forcing cone, thus allowing the bullet nose to ram its way into and through the forcing cone prior to touching the rifling. Remember that the bullet base has expanded to some degree during its passage from the cartridge case, through the chamber, through the gap, and into the forcing cone. The forcing cone swages down the entire bullet, otherwise the expanded bullet base probably could not enter and pass through the barrel.

As the bullet nose strikes the rifling, the bullet base is still entering the forcing cone. This can be a crucial situation. Rifling is placed in a barrel to force the projectile to rotate, thus imparting stability to an otherwise inaccurate projectile. Note that, when the bullets were fired out of the cylinder without a barrel, most of them turned, tipped, and went into the Duxseal sideways. Rifling prevents this under most conditions. Depending upon bullet velocity, when the bullet nose strikes the rifling, the rifling should grab the bullet and commence rotating it. However, quite often, the bullet nose will commence turning before the bullet base has entered the rifling, thus placing tremendous torque and strain upon the bullet. At this point, several things may happen. The bullet can be so disfigured that it cannot produce accuracy even though it finally does travel down the barrel. The bullet can also skid for a couple of inches before the badly stressed bullet is correctly caught by the rifling and commences to rotate. Rifling design, depth, and barrel dimensions contribute in varying degrees to what takes place. Jacketed bullets usually withstand these strains well, and the jackets permit the rifling to grab the projectile and start it rotating almost at once. Because of the jacket, disfiguration is held to a minimum in most instances. Unfortunately, lead bullets do not respond well to rifling. At the higher velocities, they sometimes will skid for some distance down the barrel because the lead bullet does not allow the rifling to grab since it is so much softer than bullet jacket material.

During the actual firing of high velocity lead bullets in the two-inch barrel, there will be instances of both barrel skidding and inaccuracy at the target.

When a handgun is fired, there must be some degree of movement known as recoil. Recoil forces the barrel muzzle to rise. Usually the lower the velocity, the less barrel movement there will be. Conversely, high-velocity heavy bullets will move the barrel muzzle substantially. On the other hand, high-velocity lightweight bullets can often produce very little muzzle jump.

The amount of muzzle jump combined with the length of time the bullet is in the barrel will drastically affect the location of the bullet impact upon the target. First, low-velocity lead bullet loads, as used in Match Target shooting, will customarily have low velocity, thus long barrel time combined with very moderate recoil. However, the long barrel time permits the muzzle to move upward, and therefore, throw the bullet high on the target. Most match bullets weigh around 146 to 150 grains, which is on the heavy side for the .38 Special and the .357 Magnum bullets. We also have a traditional Police loading that consists of the 200 grain lead bullet moving out at around 700 fps. Obviously, there is more muzzle jump with a 200 grain bullet than with a 146 grain bullet, so this particular loading usually places the bullet high on target. This is one reason why we have such high front sights on police revolvers: the high front sight forces the barrel muzzle down to place the bullet where it should go.

This same revolver with the high front sight will place 110 grain, jacketed, high-velocity bullets very low if there has been no sight change. This occurs because of the very short barrel time of the high-velocity bullet: around 1600 fps. Almost as soon as the muzzle commences to rise, the bullet leaves the muzzle, thus flying low to the target. Most of my work handguns have had the front sights cut down so that high-velocity loadings will place the bullet in the desired area of the target.

When I was in the bullet and ammunition business, one of the headaches was buying suitable bullet jackets; jackets that would allow the pure lead core to expand rapidly upon impact with the target would not stand up under the tremendous torque inherent in handgun barrels. Jackets and cores would be ripped to pieces. Jackets that would stand up under these strains would often not permit the bullet core to expand and perform correctly upon impact. As a result, bullet design had to be changed.

Far too many shooters have no understanding of bullet rotation. Assume we have a revolver with a six-inch barrel and rifling with one turn in eighteen inches. This means that for every twelve inches or one foot of bullet travel, the bullet will rotate thirty-six times. If velocity is 1000 fps, the bullet will be turning at 5066 revolutions per second, or 39,960 revolutions per minute (RPM). The higher the velocity and the faster the barrel twist, the higher the rate of RPM. Remember that we have a soft lead core encased partially within a gliding metal jacket. Remember that this jacket has been ripped and squeezed and shoved down the barrel. If we want accuracy, we must design and produce a projectile that will withstand all these forces and give us acceptable accuracy. Once the entire bullet has entered the barrel, it has been once again swaged down and resized to fit the barrel dimensions, providing of course that the jacket is reasonably malleable, which is usually not the case with full-metal-jacketed bullets. These have hard jackets and hard lead cores. On the other hand, modern jacketed bullets have tough, thick, but malleable jackets that give, stretch, and absorb the shocks and pressures encountered during the passage of the bullet down the barrel. By the time the bullet emerges from the barrel muzzle, it should be flying accurately towards the target.

Lead bullets tend to coat the barrel with lead depending upon bullet velocity, lubricant, quality of lubricant, smoothness of barrel surfaces, and other details. If leading is permitted to build up, it will so coat the barrel that accuracy will fall off badly because the lead bullets are unable to wipe out the lead and will continue to add minute particles of lead to the already leaded barrel. After a relatively short period of time, the barrel will become unable to cope with the bullets and accuracy will fall off badly. When the barrel grooves and the leading edges of the rifling become so covered with lead that a lead bullet will be torn and swaged up and then down during its passage through the barrel, it becomes difficult, if not impossible, to place the bullet into the target.

Conversely, if correctly designed jacketed bullets are fired in a barrel, even at high velocities, few, if any, problems can arise within the barrel. A copper wash will generally coat the barrel but this is so thin that it cannot affect either the barrel or the projectile. Every now and then, a barrel will get through with a tiny rough spot. This can create many problems by ripping and tearing small pieces off the bullet jacket; eventually a lump can build up that will affect accuracy.

THE EFFECT OF BARREL LENGTH UPON VELOCITY

We all know that barrels guide and direct the bullet to the target, rotating the projectile to stabilize it so that it will strike at the desired point. Most people do not realize that barrels also contribute substantially to velocity. It would be interesting to be able to fling a cartridge into the air and then fire it as it fell through the air and obtain a valid velocity reading. Since this is impractical, the next best thing to do is to fire a cartridge within a revolver chamber only and chronograph the bullet velocity.

The Dan Wesson revolver lends itself ideally to such a project; however, both the range and the chronograph had to be specially set up for this test. First, when the barrel is removed from a revolver, the bullet can fly off course due to the deformation of the base and the lack of any guidance. In addition, firing lead bullets creates problems with lead filling the barrel threads in the revolver frame. While not critical, this does produce extra work and problems. As a result, only one loading was chosen for firing: Federal #357-A, 158 grain soft-point jacketed load.

Next, due to the erratic flight of the bullet, steel plates had to be set up to protect the Oehler Skyscreens. In addition, we had to make certain that the entire area down range on both sides and above was cleared, just in case one of the bullets went completely wild. We did not wish to take any chances of injuring people or animals.

In order to help the bullets pass directly over the Oehler Skyscreens, they were located five feet apart rather than the more consistently accurate ten-foot spacing that was used throughout the balance of the tests. We estimated that we would lose 40 percent of the bullets. Surprisingly, less than 5 percent failed to give us a good reading.

The Dan Wesson revolver was set up in the Ransom Machine Rest to eliminate human error in sighting. The four-inch barrel was installed and a series of shots fired to verify previous results and to compare velocities with this ammunition with the tests run previously at the ten-foot spacing. The ten-foot spacing had given us a low figure of 1146 fps and a high of 1207 fps. With the five-foot spacing, we obtained a low of 1142 fps with a high of 1176 fps.

Removing the barrel, a dozen or more shots were fired to attempt to adjust the average bullet trajectory reasonably close to the top of the deflection plates. Once this was set up, the revolver was loaded and a series of eighteen shots fired for velocity. We lost four bullets, one of which struck the far deflection plate.

The results of the good shots are listed below:

Minimum Velocity	655 fps
Maximum Velocity	715 fps
Extreme Spread	60 fps
Average Velocity	686 fps
Standard Deviation	23 fps

Compare these figures with the velocities obtained when firing this same ammunition in the four-inch barrel:

Minimum Velocity	1146 fps
Maximum Velocity	1207 fps
Extreme spread	61 fps
Average Velocity	1183 fps
Standard Deviation	25 fps

Note that the Standard Deviation is almost identical, as are the Extreme Spread figures. The four-inch barrel appears to nearly double bullet velocity.

It should be obvious from the above results that the cartridges require a burning chamber and that the barrel provides this requirement. It would also appear that a four-inch barrel is close to ideal, because the velocity gain in moving from a four-inch barrel to a six-inch barrel is not all that large. Two-inch barrels leave a great deal to be desired. In any testing of ammunition, I strongly recommend that all three barrels lengths be used, the six-, four-, and two-inch, because far too many people fail to appreciate the variations in velocities between the barrel lengths. Because bullet performance at the target is tied closely to velocities, handgun performances must be related to barrel length.

The Additional Effect of Powder Burning Characteristics

This subject is seldom understood by the average handgun owner and shooter. However, it plays an important part in determining what the bullet will do when it strikes the target; the further the target is from the handgun muzzle, the more this subject will affect bullet performance at the target.

We have fast-burning powders, moderately burning powders, and slow-burning powders. Fast-burning powders work well in short barrels but what happens when we use a fast-burning powder and fire the cartridge in a long barrel? Will velocity increase or decrease? Conversely, if we fire a cartridge loaded with slow-burning powder in short barrels, how will this affect velocities? In order to determine the results, it is necessary to conduct a series of tests.

We require as many constants as possible. For this reason, I used the Dan Wesson revolver with its two-, four-, six-, and eight-inch barrels. Cases were Remington, once fired, .357 Magnum. Primers were Federal No. 100 Small Pistol Primers. Bullets were Hornady .38 caliber, 158 grain, .357 Flat-Point Jacketed. Loading data was taken directly from Volume Two, Hornady reloading manual using the data giving 900 fps with this specific bullet.

I considered using many powders including some of the "new" ones; however, once again bowing to the conservative thinking of the American handgunner and law enforcement officer, I turned to long-used, time-tested powders: Bullseye, Herco, and 2400.

According to the Hornady reloading manual, 6.0 grains of Bullseye, 7.5 grains of Herco, and 10.5 grains of 2400 will produce 900 fps. These are moderate, acceptable loadings and are the ones used in these tests.

Each case was inspected, then cleaned and the primer pocket cleaned after depriming. RCBS carboloy dies were used for resizing. The case mouth was then expanded in another die. Next, the powder was weighed on an Ohaus 10-10 scale and dumped into the case. Bullets were seated with another die. Finally, the case mouths were crimped into the bullet cannelure with still another die. There was no attempt to hurry the job.

Six cartridges were loaded for each barrel length with each of the three powders, requiring seventy-two cartridges for the tests. The Dan Wesson revolver was locked into the Ransom Machine Rest not because accuracy tests were being run but because each bullet had to pass over the Skyscreens in approximately the same place, thus ensuring an additional constant.

The results are given below:

BARREL LENGTH 6.0 GRAINS BULLSEYE	VELOCITY FIGURES				
2	752	786	34	767	13
4	788	838	40	807	13
6	826	871	45	851	19
	(This was an extremely accurate load)				
8	749	789	40	774	18
7.5 GRAINS HERCO					
2	880	946	66	898	24

4	908	990	82	947	26
6	973	1014	41	991	15
	(Extremely accurate load)				
8	864	965	101	930	35

10.5 GRAINS #2400

2	644	718	74	693	28
4	674	775	101	735	37
6	711	838	127	804	49
8	676	729	53	708	28

It is obvious that the above velocity figures fall short of the published 900 fps in the Hornady Handloading book. On the other hand, temperatures during this testing ran in the high forties. As a result, it would be acceptable to add an additional thirty feet per second velocity to these figures to compensate for the colder weather. Most laboratories are kept at around seventy degrees. Note that in each instance velocities fell off in the eight-inch barrel, indicating that pressures had peaked and the bullet was being slowed down due to friction and lack of gas pressure upon the bullet base.

Note also that, with both Bullseye powder and Herco, exceptionally accurate groups were recorded at fifty yards using the six-inch barrel. It would appear that this particular combination worked well.

Based upon this test, it would appear that Bullseye peaks too soon for long barrels as does Herco. The same holds true of #2400 in moderate loads.

The question now arises as to whether or not factory, full power loads would produce basically the same results. Using Remington 158 grain, flat-nose, jacketed, high-velocity loadings, the following results were obtained. Bear in mind that these are maximum loadings.

BARREL LENGTH	VELOCITY FIGURES				
2	1086	1124	38	1105	18
4	1178	1230	52	1215	20

6	1205	1267	62	1237	28
8	1237	1307	70	1264	24

The increase in velocities between the six-inch barrel and the eight-inch barrel appears to indicate that maximum velocities will be attained in an eight-inch barrel. The velocity increase is considerably smaller than normal. It would appear that the use of ten- and twelve-inch barrels calls for different powders. Otherwise, the longer barrels will probably reduce velocities if conventional handgun loads are fired.

BULLET SKIDDING WITHIN THE BARREL

Lead bullets, due to the inability of the lead to grab the rifling instantly, can skid within a barrel. The amount of skidding depends to a large degree upon velocity, the lead alloy, and the type of rifling. In many instances, skidding is not important, especially when a four- or six-inch barrel is used. These relatively long barrels counter the bullet's tendency to skid and within three or four inches will commence stabilizing the bullet.

On the other hand, skidding becomes vitally important in short barrel revolvers such as the common two-inch widely used by law enforcement personnel. The problem here is that these short barrels often cannot control the bullet because the rifling is unable to grab the bullet and fully stabilize it. The result is poor accuracy. It would appear that the higher the velocity, the longer the skid within the barrel with resulting poor accuracy at the target.

Velocity is important. Quite often, low velocity match ammunition will produce excellent accuracy in a short barrel revolver. This leads the shooter to believe that the same will hold true regardless of the ammunition used. This is not the case.

The lead alloy also contributes to skidding. Obviously, the harder the alloy, the better the rifling can grab the bullet and commence rotating it. A soft alloy can often permit the bullet to literally skid right through the two-inch barrel. When this takes place, it becomes almost impossible to hit anything over ten or fifteen feet away.

Basically, there are two types of rifling. One type has nearly equal width grooves and lands. Lands are the original barrel surfaces that remain after the grooves have been cut. The other type of rifling has narrow lands and extremely wide grooves. There are also many variations in between. Smith and Wesson tends to use equal width lands and grooves. Colt uses wide grooves and narrow lands. Because of this, I ran some tests using a two-inch barrel Colt and a two-inch barrel Smith and Wesson.

I fired Federal 158 grain, lead wadcutter ammunition and Remington 200 grain, round-nose, lead bullet .38 Special loads. Bullet recovery was made relatively simple by using paraffin blocks to slow down the bullets and then catching them in Duxseal. Two paraffin blocks, each one-inch thick, were placed

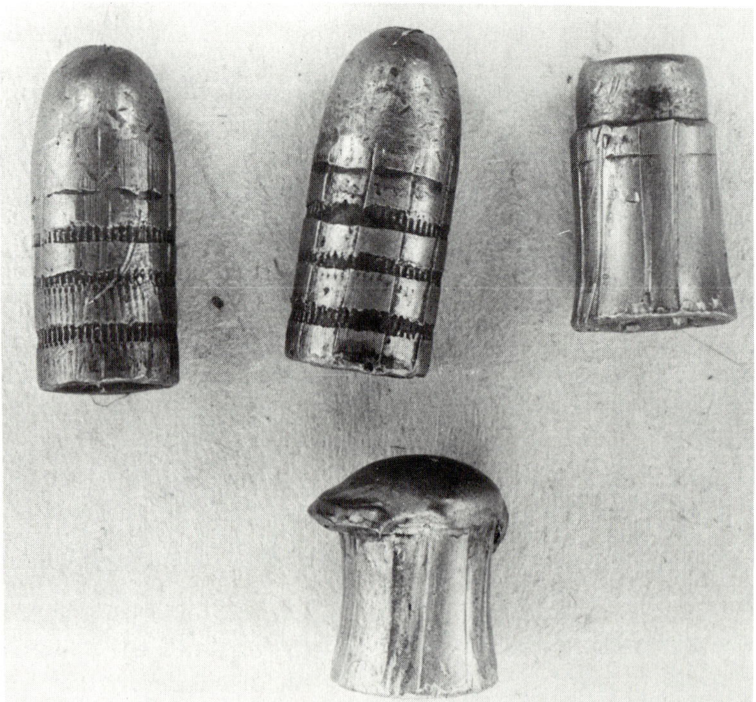

Figure 3-10. Distorted rifling marks due to bullet skidding in the barrel.

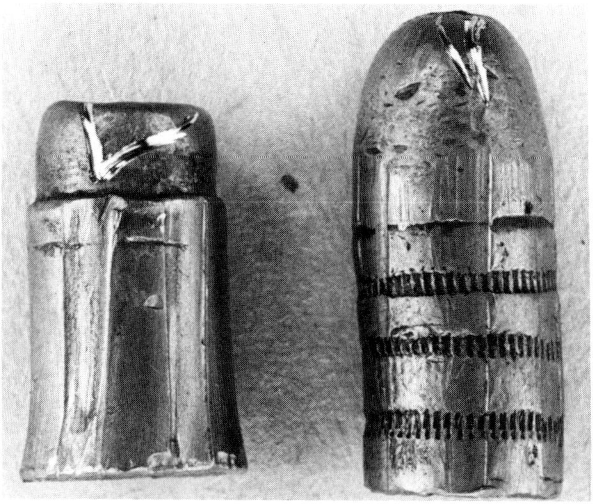

Figure 3-11. Bullet skidding is indicated by the arrows on the bullet noses.

in front of the Duxseal and the bullets were fired so that they passed through the paraffin and entered the Duxseal. This worked well, as Figures 3-10 and 3-11 show. Using the Duxseal only produced nose expansion, which I did not want. Of all the rounds fired, three illustrate perfectly the inherent tendency of a lead bullet to skid in a short barrel.

Both bullets are Remington 200 grain, lead, round-nose. One was fired out of the Colt and the other out of the Smith and Wesson. The third bullet shown with the partially expanded nose is a Federal 158 grain, .357 Magnum bullet. It is interesting in that it carries two perfectly marked land markings out of the Colt but the other land markings show the large amount of skid distortion and are much wider. The two Remington bullets each show the tendency to skid and the markings are basically alike.

It would appear that lead bullets are not suitable for use in two-inch barrel handguns if the velocity is high.

THE FOULING OF REVOLVER BARRELS

One test conducted was to demonstrate the varying velocities that may be expected from firing the same ammunition in four different makes of revolvers. A side result that was far more severe than expected was the fouling and leading of the barrels. This, in turn, stimulated a great deal of thought and discussion.

The question continually arose as to what effect, if any, this fouling would have on accuracy and on velocity. There has been a long-standing assumption that fouling of any kind would definitely affect accuracy. Because of this, I conducted the following test.

Because the average person tends to reject new ideas, I decided to discard the Dan Wesson revolver for this test and turn to the conventional and long-accepted Smith and Wesson, Heavy Barrel, Model 13-1 four-inch barrel revolver. This handgun is fitted with standard "Magna" factory grips. It came to me directly from the factory, factor test-fired only. It was used in the four revolver test. Otherwise, it has not been fired. It may be considered as a new handgun.

The small "Magna" grips were removed and the revolver locked into my Ransom Machine Rest. Eighteen rounds were fired to shoot in and set the rest. The barrel was then thoroughly cleaned, wiped dry, and checked.

My Oehler #33 Chronotach was set up with the Skyscreens spaced at ten feet. The distance from the front of the bench rest to the target faces was fifty yards.

I wanted to obtain both group sizes and velocity data of a minimum of six six-shot groups in an attempt to discover the influence of barrel fouling upon group size and velocities. Three groups were fired and the results are shown below.

Control Group Number 1

Ammunition: Remington Index #R357M2 158 grain jacketed, hollow-point, high-velocity.

Group Size (inches)		Velocity Figures				
1.	4 1/2	1204	1258	54	1243	20
2.	6 3/4	1147	1265	118	1198	44

3.	5 3/8	1136	1224	88	1182	31
4.	2 5/8	1162	1241	79	1205	32
5.	5 1/8	1155	1213	58	1190	20
6.	3 7/8	1173	1236	63	1204	23

Each group consisted of six shots.
Maximum Average Velocity: 1243 fps.
Minimum Average Velocity: 1182 fps.
Difference: 61 fps.
Thirty-six shot group size: 7 inch.

Very little fouling or lead build up. Barrel cleaned at end of thirty-six shot string.

Group Number 2

Ammunition: Federal Index Number 357C 158 grain, Semi-wadcutter, lead, high-velocity.

Group Size (inches)		Velocity Figures				
1.	4	1193	1239	46	1220	15
2.	4 3/4	1211	1235	24	1222	9
3.	8 5/8	1219	1264	45	1239	19
4.	5	1206	1257	51	1231	18
5.	10 5/8	1212	1265	53	1238	17
6.	4 3/4	1198	1244	46	1229	16

Each group consisted of six shots.
Maximum Average Velocity: 1239 fps.
Minimum Average Velocity: 1220 fps.
Difference: 19 fps.
Forty-two shot group size 11 inch.
Substantial lead buildup in barrel.

Group Number 3

Ammunition: Remington Index Number #R357M5 158 grain, lead, wadcutter, high-velocity.

Group Size (inches)		Velocity Figures				
1.	4	1177	1224	47	1204	20
2.	4 3/8	1178	1289	111	1211	41
3.	3 3/4	1182	1230	48	1199	17
4.	3 5/8	1172	1236	64	1204	23
5.	5 3/8	1184	1240	56	1207	23
6.	6 1/4	1181	1244	63	1198	23

Each group consisted of six shots.
Maximum Average Velocity: 1211 fps.
Minimum Average Velocity: 1198 fps.
Difference: 13 fps
Thirty-six shot group size 7 1/4 inch.
Substantial lead buildup in barrel.

The results of this firing prove to be rather interesting. The control Group Number 1 was fired to provide a guide to judging the other groups. I had expected that this would be a relatively tight group. I selected the Remington ammunition for no particular reason; it was the first box I picked up, so I used it.

Note that the maximum and the minimum average velocity figures showed an acceptable velocity difference of 61 fps. Those people who demand less than that are unreasonable, in my opinion. Group size also met my expectations. On paper this is an excellent fifty-yard .357 Magnum loading.

Note that there was very little fouling in the barrel and almost no unburned powder, thus appearing to be a desirable loading.

Group Number 2 turned out to be a surprise for many reasons. Usually such a hot loading firing a lead bullet load will develop unacceptable leading. While there was substantial barrel leading, it was not unreasonable.

Note that this lead bullet loading produced two "large" groups, #3 and #5, due to flyers that could be attributable to these bullets "cleaning out" the barrel to some extent and thereby becoming unstable. This did not occur with the jacketed bullet loading #R357M2. I was extremely surprised at the consistency of the velocity figures that developed a mere 19 fps. difference between maximum and minimum velocities.

The real surprise is Group Number 3, firing Remington's #R357M5 loading. The velocity difference is 13 fps. for a full thirty-six rounds. This, in my opinion, is remarkable, to say the least. In addition, group size is only one-fourth inch greater than that of the jacketed bullet loading #R357M2. Also note the lack of any flyers in any of the groups.

The buildup of fouling within the barrel continued but obviously did not affect accuracy or velocities.

Seventy-eight rounds were fired without cleaning the barrel. On close examination it appeared that barrel condition was considerably better than it had been when the four-revolver test was fired, thus bringing up the obvious question of why. Remember that only lead bullets had been fired, seventy-eight of them, and it would appear that these bullets might partially clean out or wipe out some of the leading from time to time. Do not misunderstand me — the barrel would never pass inspection in any law enforcement agency nor would any professional handgunner accept a barrel in this condition.

As this situation was discussed, we decided to take the revolver as it was and fire six rounds from every box of ammunition that we had on hand for the tests. After removal from the Ransom Rest, the "Magna" grips were installed. The revolver then existed precisely as shipped by the factory.

I then personally loaded and fired the scheduled six rounds from each box of ammunition. At this point, I would like to emphasize that I fire from 500 to 1500 rounds every couple of weeks throughout the year. I am constantly firing. I wear the big Ithaca ear muffs for anything more potent than wadcutter ammunition. All of my personal handguns have special grips, either made for me to my specifications or altered (factory grips) so that they fit my hands. I have been doing this for well over forty years, so that firing anything is no big thrill.

I wished to emphasize this because, throughout the tests using both freehand shooting and firing in the Ransom Machine Rest, I have discovered that I could not shoot continuously. I had to stop and rest due to muzzle blast, plus powder residue particles rebounding from the blast shield onto my hands, face, neck. The concussion alone is extremely tiring.

Now that I was firing six rounds with one and two hands, many negative conclusions came to mind. First, the Smith and Wesson "Magna" grips are utterly unsuitable to a .357 Magnum revolver. It is impossible to control the handgun. Further, recoil hurt the web of my hand.

Next, the recoil is so severe with *all* loadings that recovery time is slow. This is especially true of firing with one hand. A two-handed hold provides slightly better control.

The muzzle blast is entirely unacceptable even with ear muffs. Unfortunately, I have never seen anyone walking around wearing ear muffs or even carrying them while on duty. It should be obvious that no officer can be expected to draw and fire six controlled, aimed shots with any of this ammunition anywhere, particularly within a confined space such as a room or hallway.

It is no wonder that officers fire six shots and hit absolutely nothing. I have long advocated that, at the end of every training session, all involved officers be forced to fire their old issue ammunition, both in the handgun and on their belts, without ear muffs. This is their situation on the street, and it is *not right* to train men with wadcutter ammunition and then put them out on the street with .357 Magnum loadings and expect them to hit anything.

I admit it would be impractical, if not impossible, to fire eighteen fast rounds on a training course with .357 Magnum service ammunition due to the excessive heat generated by the cartridges — the handgun would be nearly impossible to touch. On the other hand, personnel should still be forced to fire their eighteen old rounds on a special course without ear muffs or other aids not available to them on the street. I believe that, when an officer draws and fires on the street, his misses are due to his shock and reaction to the muzzle blast and to the noise.

While this is not a normal ballistic subject, I believe it should be examined in depth by any reader of this book. Officers may be given any ammunition, but if they cannot hit with it, what good is it? I am interested in a more practical loading for law enforcement work — a loading that will provide possibly even more impact shock than our current super hot ammunition, and yet will not penetrate excessively and will still do the job at normal law enforcement distances. We do not need high-velocity hunting ammunition, which is what is being turned out by factories today, for both .38 Special revolvers and .357 Magnum revolvers. This entire subject should be investigated in depth by people who have had experience, not by political appointees.

Now that I have discussed muzzle blast, excessive recoil, and poor grip design, I would like to examine the barrel fouling situation. At the end of the firing of the six shots with each type of ammunition, the barrel was, once again, coated with lead and fouling. Around the muzzle on the outside was a ring of lead. A dry, clean patch could be run through the barrel with difficulty. It did not remove any of the fouling.

Basically, it would appear that if only lead bullets are fired leading will build up to a certain degree. Beyond this point, lead buildup appears to be very slow.

If jacketed bullets only are fired, a definite copper wash appears. This does not affect anything, but the moment fouling commences to build up, it becomes serious.

It would appear that, when a combination of lead bullet loads and jacketed bullet loads are fired, the combination definitely creates a buildup of lead and fouling within the barrel. Part of this could be due to powder residue from the 110 and 125 grain, jacketed bullet loads. Obviously, there is a substantial amount of unburned powder that mixes with the hot lead and the jacket wash to create a very bad situation within the barrel.

This barrel fouling cannot be wiped out with a few patches soaked in Hoppe's, nor with a brass brush. My experience has been that all this does is "wash and clean" the barrel fouling. I had to fall back upon both Bobst Bore Cleaning Compound and very fine valve grinding compound, used gently and sparingly, along with heavily soaked patches. In my opinion, this is an intolerable situation and one that should be corrected rapidly. Both the handgun manufacturers and the ammunition manufacturers should get together and work out a practical solution to this problem.

Chapter 4

THE SEMI-AUTOMATIC PISTOL

WE have discussed in some detail the characteristics and problems inherent in revolvers. I would now like to do the same with the pistol and its self-functioning mechanism. In a revolver, the shooter must manually activate the mechanism. The pistol, on the other hand, requires only that the shooter press the trigger, thus commencing a complete cycle of functioning carried on solely by the cartridge and the pistol's mechanism.

Pulling the trigger allows the firing pin of the pistol to fall and strike the cartridge primer, thus firing the cartridge. The powder ignites, and pressures rise. Finally, the bullet commences to move out of the cartridge case mouth, down the pistol barrel, and out the muzzle.

Pistols must be separated into two design classes — those with locked breeches and those with straight blow back mechanisms. Basically, most, if not all, 9 mm, .38 Super, and .45 ACP pistols contain a breech locking mechanism that prevents the breech from moving to the rear until pressures have fallen substantially. In most instances, the barrel and breech move to the rear together for a short distance, permitting the bullet to leave the pistol barrel muzzle. Then, an unlocking mechanism takes over and the breech continues its movement to the rear, leaving the barrel ready to be rammed forward again into its forward position.

During its rearward movement, the breech, by means of an extractor incorporated into the breech face, pulls out the fired cartridge case and brings it back until it strikes the ejector that rams the empty case out through the ejector port.

The slide continues its rearward movement until it is well to the rear of the magazine. It then stops, moves forward, and strips the top cartridge in the magazine forward and out of the magazine. The cartridge moves forward and slightly upward until the bullet nose strikes a feeding ramp of some kind that guides the bullet nose into the barrel chamber. The slide then rams the fresh cartridge home into the barrel chamber, shoves the barrel all the way forward into battery, and the pistol is ready to be fired again. All this is done mechanically without the shooter having to do anything more than release the trigger. The semi-automatic pistol is prevented from firing a second shot automatically by a disconnector that functions in connection with a cut-out or cam, or some other method that keeps a part of the trigger assembly disengaged until the pistol is back in battery and the trigger released. Once the trigger has been released, the pistol may again be fired.

A pistol cartridge lies within a chamber that is part of the pistol barrel, thus eliminating the cylinder that is a part of the revolver. This does away with bullet jump from the cylinder into the barrel throat. There is nothing to line up correctly prior to firing. Basically, a pistol is far more rugged than a revolver.

Pistol cartridges are pressed into a magazine against a spring that forces the cartridges up against the top lips of the magazine so that a single cartridge is always ready to be rammed into the barrel chamber by the slide. When the last cartridge has been fed out of the magazine into the chamber, the follower (on which the bottom cartridge rests) rises into the path of the bottom of the breech face that is then prevented from moving forward, thus locking the slide back indicating to the shooter that the pistol is empty.

The locked breech type (that holds the breech face and the barrel tightly together for a brief period of time after firing) of mechanism is necessary in the more powerful calibers to prevent the slide from ramming back under high pressures and eventually damaging the pistol mechanism. The second type of design is the straight blow back that is capable of handling cartridges like the .22 long rifle, .32 ACP, .25 ACP, .380 ACP. These smaller cartridges may be held safely by the breech block due to its designed ability to provide sufficient inertia to handle these cartridges. Pistols such as the Walther PP and PPK are all blow back designs.

Every now and then a pistol crops up that has been converted from locked breech to straight blow back. Many years ago, the

Army Marksmanship Unit, and later many gunsmiths, altered the Colt .38 Super to handle the .38 Special, match, wadcutter cartridge. This cartridge had the bullet seated flush with the end of the cartridge case and was loaded to very low, match pressures. Accuracy was excellent, and for a time, this conversion was rather widely used by professional match shooters. Pressures were so low that the blow back design was perfectly adequate and safe. Such a pistol should never be used to fire any ammunition except match wadcutter cartridges.

When a pistol cartridge fires, it is completely enclosed within a chamber that is a part of the pistol barrel, thus the bullet moves forward to instantly contact the barrel rifling. There is no gap to jump nor any potential problems awaiting the bullet. As a result, the pistol bullet leaves the muzzle in far better condition than the average revolver bullet. Such a design also eliminates having six individual chambers. It provides an almost perfect gas seal, thus allowing the pistol to fully utilize the full potentials inherent in the powder charge. In this connection, pistol cartridges are far more efficient than revolver cartridges because they have had the full advantages of better powders with resulting more practical cartridge design. No pistol cartridge has ever been a black powder cartridge as is the case with revolver cartridges. Pistol cartridges have been designed around high intensity, smokeless powders from the start. There are no holdovers from black powder days.

Pistol mechanisms do have problem potentials. They were designed to function with full-metal-jacketed bullets; when cartridges containing these bullets are used, there are seldom any problems, providing of course, that factory cartridges are used. The moment a handloader becomes involved with a pistol, he usually turns to lead bullets, which are inexpensive, low-velocity loads, with other ammunition variables. This can be one source of problems but the competent handloader has learned how to compensate for his ammunition changes.

Few professional handgunners can keep from modifying pistols to suit their personal preferences. If the slide is lightened, problems arise. If something else is altered, then problems arise. The pistol shooter must realize ahead of time that any change in the pistol or ammunition components must be accompanied by a corresponding change somewhere else. Otherwise, functioning, accuracy, and pistol life can be seriously affected. If a shooter wishes to use low-velocity, handloaded ammunition in a pistol, the odds are that he will be forced to alter the recoil spring to permit the pistol slide to be rammed all the way back to allow it to strip off the top cartridge in the magazine. The spring must retain sufficient force to ram the slide forward hard enough to strip off the cartridge, ram it into the chamber, and bring the barrel forward into battery. All these details must be worked out for each specific pistol.

The pistol has several design advantages over the revolver in other areas. The student of firearms knows that, during the original acceptance tests by Army Ordinance, the Model 1911 was subjected to dirt, dust, and mud tests that were brutal. The design of this pistol helps prevent foreign matter from entering the pistol mechanism. I have personally taken this pistol, put cellophane tape over the muzzle and thrown it into a bucket of mud, sloshed in more dirt, sand, and water, turned the whole mess upside down on the ground, and gone off and left it for half an hour. Returning, I removed the pistol from the dirt and fired the entire magazine. I have watched a platoon of Canadian Army personnel bury five Browning/Inglis Model 1935 pistols on a parade field, march up and down for a while over these pistols, dig them out, and then fire each one sixteen rounds without malfunction. Obviously, the larger the caliber, the more reliable a pistol will be when subjected to treatment of this type; this is one reason for the inherent reliability of the Colt Model 1911 pistol.

Revolvers cannot take this abuse due to their open construction and the necessity of having relatively delicate parts turn the cylinder in order to fire the handgun. I do not recommend that anyone attempt to duplicate the above tests with a revolver because it would be extremely dangerous — due to the six chambers with their open ends plus the barrel throat, plus the barrel muzzle even though the latter is covered with cellophane tape. Another advantage of the pistol over the revolver is that it contains no cranes, ejector rods, or exposed barrel that may be bent. The pistol will take a lot of abuse and continue to function; revolvers will not.

I do not wish the reader to believe that I am being overly critical of revolvers. Pistols do have their problems, such as extractors that break and magazines that become defective due to long use or excessive abuse. Basically, pistols tend to be far more reliable for longer periods of time than revolvers.

Chapter 5

IDENTIFICATION OF HANDGUN AND AMMUNITION COMPONENTS

As anyone who reads the comic strip "Dick Tracy" or delves into the subtleties of murder mysteries knows, the officer or the hero must always find the fatal bullet so that he can match it to the barrel of the murder weapon. He can thus pin the crime on the suitable villain. Up to a point, this is true and the relationship between the barrel characteristics and the markings on the bullet should be established. Each barrel has tiny, individualistic marks that distinguish it from any other barrel. These marks are put into the steel when the rifling is cut, swaged, or forged into the barrel steel. Such an identification is not always possible due to the controls over rifling methods and the care with which quality control supervises the final finishing of the barrel interiors. Many years ago, the FBI complained to Smith and Wesson that their rifling procedures were so perfect that in many instances it had been impossible to identify and match a specific bullet to a specific barrel.

The ability to match can vary. Assume that a homicide is committed using a specific handgun, specifically a revolver, and that the bullet is recovered. After suitable inspection, it is determined that this specific bullet had passed through this specific revolver barrel. The criminal is then located and punished. However, suppose that, after firing this bullet, the criminal goes somewhere and fires thirty to fifty cartridges as rapidly as possible through this revolver, alternating lead bullet loadings with high-velocity jacketed bullet loadings using perhaps Remington, Federal, and Winchester ammunition. The barrel is not cleaned. The revolver is then found by the police. Can they match the barrel now to the fired murder bullet? The odds are they cannot get positive identification. Barrels change slightly every time the hot gases, the lead, and the gilding metal jackets pass through at high velocities. Or, assume that immediately upon firing the fatal shot, the criminal picks up a small, rat-tail file and runs it through the revolver barrel several times, then throws away the file. Matching the bullet to the barrel becomes impossible.

If we have a criminal of better than average intelligence, after firing the fatal bullet, he would then remove the barrel and replace it with either a new barrel or an old barrel, thus presenting the police with an entirely new set of problems. The murder barrel can be packed with salt, wrapped in old newspapers, wet down thoroughly, and casually planted in a flower bed or at the base of a shrub or thrown into a river. Even if found, within a matter of days such a barrel would become so rusted that matching a bullet to it would be impossible.

An even more calculating criminal would remove the barrel of the revolver before committing the homicide, shove the revolver frame into the victim's body and then fire, return to his base of operations and install the barrel once again on the revolver frame. Police would have no means of matching the bullet to the barrel because the barrel had not been used to project the bullet into the victim.

As may be seen, under certain circumstances, a fatal bullet cannot be matched to a specific handgun.

The bullet is merely one of several cartridge components. Assuming that the bullet cannot be matched to the barrel, the next place to turn is to the cartridge case. When the cartridge fires, the cartridge case expands and, for a few milliseconds, is jammed tightly against the walls of the cylinder chamber. Sometimes it is possible to locate specific scratches or marks on the cartridge case and then match them to corresponding marks inside one of the chambers. By firing cartridges similar to the murder cartridge in the revolver cylinder, marks on the cartridge case can be matched to the murder weapon. It is important in every case that identical ammunition to that used in the homicide be used, otherwise results can vary substantially.

Directly behind the top of the revolver cylinder is the recoil plate. This plate is usually inserted during assembly of the handgun and is designed to take the full thrust of the cartridge when it is fired. This plate supports the head of the cartridge case and is generally slightly forward of the balance of the rest of the frame at the rear of the cylinder. By moving the recoil plate slightly forward, it will more readily support the base or head of the cartridge case. Upon firing the cartridge, the case tends to move back solidly against the recoil plate at the same time that the cartridge case walls expand to grip the chamber walls. Quite often, marks may be found on the head of the fired case that match the markings of the recoil plate. Another very individualistic set of marks may be found where the firing pin protrudes through the center of the recoil plate. If a revolver has been dry fired to any extent, the lips of the small hole can become pushed out. Running the finger over the recoil plate will allow a person to feel the slight bulge around the firing pin hole. All these markings can often be found on the base of the cartridge case and on the primer.

Similarly, the firing pin itself can have individualistic markings such as tiny burrs, or a slightly offset point, or too sharp a point or too blunt a point. These indentations of the firing pin nose in the primer of a fired case can definitely aid in identifying the revolver that fired the bullet.

I believe it would be wise at this point to emphasize that, with the tremendous growth of handgunning in this country in recent years, more and more shooters are no longer satisfied with factory workmanship. They stone, smooth, and work over each individual part or have master gunsmiths (like Ron Power or Jim Clark) literally rebuild their handguns. These worked-over handguns are instantly recognizable and stand apart from standard factory pieces. The problem is that there is nothing to stop the criminal from also stoning and polishing the identifying parts of his handgun, thus destroying any chance of matching up firing pin indentations or recoil plate markings. A professional handgunner can so alter his handgun after firing a single shot that it can never be tied in with that bullet. Fortunately, the average criminal seldom has the knowledge to go the whole route and is, therefore, caught on some technicality or detail that he overlooked.

Insofar as automatic pistols go, barrels may be rapidly replaced in many instances. Breech block faces may be stoned and firing pin noses altered and even shortened or replaced with old ones or new ones, thus giving entirely different firing pin indentations in fired primers. Usually, parts may be exchanged and taken from other similar pistols so the resulting hodge podge not only functions but will completely confuse ballisticians. Matching bullets or fired cases to a specific pistol can become impossible.

Once again, I wish to stress that handgunners smooth and polish pistols so that functioning and reliability are more constant during competition; the officer or lab technician should not hassle a "gun nut" or competition shooter just because the pistol has been smoothed and hand finished. Once again, there is nothing to prevent the criminal from doing this; however, generally speaking, the criminal does not understand what to do.

There is an entirely different phase to all this that is not usually understood. As an example, how many people know that the Remington .350 Magnum rifle can fire a .38 Special bullet? This is one of the joys of owning this remarkably light, handy little rifle. It may be used with lead bullet loads or with jacketed bullet loads, both of which produce little to no recoil and ram out the handgun bullets at velocities around 2,000 fps., thus giving the shooter an excellent varmint and small game rifle and cartridge. Suppose an officer finds a .38 caliber bullet at the scene of the crime with a couple more in the victim's body, what should he do? The suspect does not own a handgun. He has never owned a handgun. But he does own a .350 Remington Magnum and handloads for it.

Today the ability to fire handgun cartridges or bullets in many different rifles and calibers can affect an investigation. Officers should be trained to notice details such as a part-time machine shop or a reloading bench, however small and compact with its dies and tools. Cartridge adapters are commonplace today so that even if the suspect does not reload nor have a machine shop as a hobby, he can still purchase adapters and commit a murder with a rifle and practically force the police to assume that a handgun was used. If barrel markings on a bullet do not match conventional handgun barrel rifling, then, by all means, check out the suspect's dwelling for rifles of the same caliber as the murder bullet. I am speaking of the whole range of interchangable calibers — .22s, .25s, .32s, and on up. Marlin and Ruger produce carbines that fire the .44 Magnum cartridge.

Winchester's Model 94 rifle has been altered to take .357 Magnum and .44 Magnum cartridges. These variations should be carefully looked into.

Adapters are available for the following:

.22LR in .220 Swift	.22 Hornet in .220 Swift
.22LR in .221 Fireball	.22 Hornet in .22-250
.22LR in .222 Rem	.30M1 Carbine in .30-30
.22LR in .222 Rem Mag	.30M1 Carbine in .30-06
.22LR in .223 Rem	.30M1 Carbine in .308 Win
.22LR in .22-250	.30M1 Carbine in .300 Win. M
.22WMR in .220 Swift	.32ACP in .30-30
.22WMR in .222 Rem	.32ACP in .300 Win. M
.22WMR in .223 Rem	.32ACP in .30-06
.22WMR in .22-250	.32ACP in .308

Special adapters for Thompson Contender:

.22LR in .22 Hornet TC	.22LR in .222 Rem TC
.22LR in .221 Fireball TC	.22WMR in .222 Rem TC
.22WMR in .221 Fireball TC	.22WMR in .22 Jet TC
.22LR in .223 Rem TC	.22LR in .218 Bee TC
.22WMR in .223 Rem TC	.22WMR in .218 Bee TC
.22LR in .22 Jet TC	.38 SPL in .357 Herrett

It is not enough to check out factory rifling dimensions or even handgun barrel rifling dimensions because of the large number of rifle barrel makers turning out barrels today. Many of these barrels may be used for both handgun and rifle. Many custom gunsmiths specify special barrel rifling configurations or special twists that are not generally used but that may be ideal for firing a specific bullet out of a special barrel for a special match. All these variations pose problems for police laboratories and they should approach these problems with an open mind and a desire to discuss and evaluate them with gunsmiths and professional shooters. Laboratory technicians can no longer act as if they alone hold the answers to all problems.

I have never yet met a professional shooter or competent

gunsmith who would not sit down and talk with police officers and lab technicians. Most American shooters are law-abiding individuals who have been maligned and abused beyond all reason. Even today, far too many law enforcement personnel consider the American shooter as a moron. The police often fail to appreciate the tremendous body of shooting information contained in the minds of a group of match shooters. Never forget these men and women have chosen to enter handgun shooting. They enjoy it. They have experimented and tested and evaluated to attempt to give themselves the finest possible competitive handgun. This is not true of the average officers or lab technicians, who far too often are *assigned* the job, not because they like guns, ammunition, and ballistics and own perhaps several dozen with which they are constantly tinkering. Vast potential lies untapped by law enforcement, and it is time that law enforcement commences to work with the civilian shooter and gunsmith.

HANDGUN AND RIFLE BARRELS

There appear to be many inconsistencies when we attempt to compare handgun barrels with rifle barrels. As an example, in caliber .30-06, the 150 grain and the 180 grain bullets are widely used. Because of this, a rifling twist of one turn in ten inches is used, thus correctly stabilizing the bullets at their normal velocities. If a faster twist is used, such as one turn in eight inches, the bullets may be stabilized at lower than normal velocities. Conversely, if a slower twist is used, then the bullets must have a considerably faster velocity in order to become stabilized. This latter situation is often impractical because in order to produce higher velocities, chamber pressures may be so high that acceptable pressures will be exceeded. In any event, bullet stabilization is a must in order to obtain acceptable accuracy at the target. In other words, there are controlling limits within which a bullet must rotate in order to fly accurately from the rifle muzzle to the target.

When we examine handgun barrel twists, we discover that in the .357 Magnum, barrel twists can run from one in fourteen inches up through one in eighteen and three-fourths inches.

This does not make sense because one weight of bullet cannot correctly be stabilized in such a variation of barrel twists. In rifles, barrel twists have customarily been related basically to the sectional density of the bullets to be fired in that particular rifle. A bullet weighing 180 grains in .30-06 caliber with a sectional density of 0.270 would require a one in ten twist. A bullet weighing 110 grains in .30-06 caliber with a sectional density of 0.166 would need a barrel twist of one in fourteen. This is not used in handgun barrels.

The sectional density of .357 Magnum bullets is relatively low, running in the low to middle 0.100. Put still another way, a long heavy bullet needs a fast twist but most handgun bullets are short, fat, and chunky thus, theoretically, requiring a slow twist.

It is interesting to note that around the turn of the century rifle makers were forced to incorporate new concepts of barrel dimensions and rifling designs in order to provide acceptable accuracy with the then new high-velocity, small-bore jacketed bullets. Examine the Springfield, the Enfield, the Ruger, the Remington, the Winchester, the M-14 and you will discover that high power rifles today utilize a wide groove and narrow land rifling concept in order to correctly handle the jacketed bullets. No longer do we find the shallow, wide groove, wide land rifling that was in common use prior to the advent of the high-velocity, jacketed, small-bore bullet.

One of the problems that arose with the introduction of long-jacketed handgun bullets was the skidding and jacket stripping that occurred when they were fired in conventional lead bullet rifling. If the handgun barrel was short, accuracy would suffer. Usually in a four- or six-inch barrel there was sufficient barrel time for the bullet to stabilize and to commence rotating correctly. Even today, with the jacketed bullet an accepted handgun design, we find that some handgun barrels continue to retain the classic lead bullet rifling conformation of wide grooves and wide lands. Others are moving towards wide grooves and narrow lands. Some rifling is shallow. Some rifling is deep.

Bullet stabilization is the ultimate goal of barrel twist. This stabilization should be achieved without deforming the bullet or the jacket, which means that the correct type of rifling must be used. Returning to the concept that a high-velocity bullet requires less twist than a slower moving heavier bullet, let's use this as a base of thinking. A revolver that will fire a 110 grain .357 Magnum bullet should have a twist of perhaps one in twenty. A 158 grain .357 Magnum bullet should then have a barrel twist of perhaps one in twelve. As it is, there is no barrel twist compensation for varying bullet weights in handgun barrels as they are now commercially produced. Possibly, it is not immediately feasible.

If we cannot expect barrel twists to match bullet weight, then we should be able to expect that barrels will be rifled to correctly handle bullet types such as wide grooves and narrow lands for high-velocity, jacketed bullet loadings and shallow, equal width grooves and lands for slow-moving lead bullet target loadings. A barrel that will produce gilt edge accuracy with a 146 grain, wadcutter, lead bullet should be relatively inaccurate when high-velocity jacketed bullets are fired in this barrel.

Today, we have many fine barrel makers plus an increasing number of extremely competent, capable gunsmiths who know from personal experience the demands and requirements of competitive handgun shooting. Handgun shooting today is a far cry from the traditional NRA and USRA matches that make up only a small portion of the total competitive handgun matches being fired. These barrel makers and gunsmiths have forced the shooter to take a new, hard look at handguns and their barrels. They have found that commercially produced handgun/ammunition combinations give adequate accuracy but fail to meet the standards that present day competition demands. The new breed of handgunner must have accuracy that rivals that of rifles. We have a new phase of handgunning today known as the Mexican Silhouette. This match requires accuracy and power far in excess of anything previously required in handgun competition. In addition, we also have the Laramie Match (hosted by Doctor Burgess) that places a premium upon long range accuracy. The resulting handgun potentials are gradually commencing to be realized. Commercial barrels are being replaced by custom barrels. The hand fitting of the component handgun parts to meet the extremely exacting specifications of these new matches is now big business.

Back in the 1870s, we had a type of rifling and barrel contour that proved superior to what was commonly used in those days. This was gain twist rifling combined with a slightly choked barrel. It is difficult to understand why this concept has not been adapted to today's long range match handguns. Bullet deformation is held to a minimum with this combination. I believe that the cost of producing gain twist along with a slightly choked barrel has dampened any enthusiasm handgun manufacturers might have about turning out these barrels. However, the advantages are so obvious that a slightly higher retail price should not hinder sales. Gain twist rifling eliminates any problems of bullet weight (sectional density) versus rifling twist. The slight choke nearly eliminates any bullet deformations that might take place. On paper, it appears the ideal solution.

It will be interesting to watch what happens in the next few years. It could be that custom barrel makers and gunsmiths will combine with the shooters to force the factories to sell handguns without barrels or else the large manufacturers will have to fall into line and meet the handgunner's requirements of handgun fitting and handgun accuracy. In my opinion, the accuracy potential of the handgun has not even been scratched.

Chapter 6

TEST CONDITIONS

THE BENCH REST

FAR too many things are taken for granted in testing and evaluation by those who enter this field for the first time. The basic reason we have a bench rest is to help eliminate the human error. Once a bench rest is constructed, we can then affix mechanical equipment to the top to further aid in eliminating the human error. Thirty years ago when Bench Rest Shooting commenced as an active rifle sport in this country, it was thought that bench rests could be put together using many exotic concepts of design and construction. This was not true and it is not true today. The simpler the bench rest, the more practical it will be.

A bench rest for testing and evaluation is a different project from a bench rest designed for bench rest rifle shooting. The rifle shooting bench rest is not required to take recoil. The type of bench rest we are discussing must be so designed and constructed that it will handle shock and recoil.

Normally, when one thinks of solidity and strength, reinforced concrete comes to mind. This type of construction should not be used for the bench rest. There is no give or shock absorption to concrete. Possibly the finest material available is railroad ties. Four eight-foot railroad ties sunk four feet into the ground and topped by other railroad ties and a top platform of more railroad ties is ideal. Such massive construction is generally unnecessary, too expensive, and too time-consuming. I definitely recommend railroad ties for the posts that go into the ground. The main beams that tie the four posts together should be at least two by sixes or two by eights. Over these should be laid two by sixes to form a solid, massive bench rest top. Such construction will handle any handgun testing.

The reason for the use of wood throughout is that wood will accept, absorb, and handle shock. Concrete will handle shock

Figure 6-1. A solid bench rest, suitable both for chronographing and for firing handguns in a machine rest.

Figure 6-2. A heavy, carefully constructed bench rest, suitable for bench rest rifle shooting, long range handgun and ammunition testing, and firing handguns in machine rests at long ranges. Note the moveable seats.

but will not absorb it. When we fire a handgun in a machine rest, we cannot have double shock or lack of shock absorption. Double shock is failure of the bench rest to accept or absorb shock, thus sending it violently back into the firing mechanism and affecting the validity of the test machine.

The simpler the construction, the better, I know of no reason to complicate the design or construction of a bench rest. It is important that all lug bolts and other bolts be kept drawn up tight. Nails must be solidly set. If the main supports become loose in the ground, they must be reset. The entire bench rest must be solidly constructed and kept that way, otherwise, the validity of the tests can be questioned.

THE HANDGUN MACHINE REST

It is impossible to efficiently or effectively evaluate handgun or ammunition potential without using a machine rest. There are many kinds. I have found, over the years, that one that basically reproduces a hand-held hold is superior to those types that utilize only recoil, V rests, or parallel bars. A functioning handgun whether revolver or semiautomatic pistol or single shot pistol produces empirical results only when fired under conditions similar to actual firing conditions. Artificial test conditions may prove specifics but they will not reproduce actual firing situations. I have a very fine ammunition firing mechanism that will tell me precisely what .38 Special ammunition will do in that device but these results cannot relate to firing the same ammunition in a hand-held handgun.

A correctly designed and created machine rest will just about duplicate results obtained through hand-held firing. By this, I mean that if the machine rest, the handgun, and the ammunition will turn in two-inch, ten-shot groups at fifty yards consistently, when this handgun/ammunition combination is hand-held and fired it too will produce two-inch ten-shot groups at fifty yards. There should be and can be duplication of results. This is the purpose of a machine rest.

A machine rest should be used for evaluation and testing to eliminate the human error. With today's high velocity ammunition, it is doubtful if long strings of fire can be shot by anyone with consistent results that correctly evaluate handgun/ammunition accuracy. A machine rest will do this all day long. It is tiring enough to work with high velocity ammunition in a machine rest without attempting to duplicate this through hand-held firing.

Over the years, I have used many various kinds of machine rests. Today, I believe that the Ransom Rest is possibly the most reliable handgun machine rest on the market. It is extremely versatile and will take just about any handgun through use of removable "grips." It permits recoil and hand-held motion. This mechanism operates through use of a slip clutch principle. The pressure on the contact plates may be varied. It is simple and rugged.

This mechanism is screwed onto a sheet of one-inch plywood. I use the finest Marine quality plywood I can buy. I have painted the plywood with several coats of good outside paint. This plywood is then screwed onto one by twos running across the bottom ends of the sheet about two to three inches in from the ends. Holes have been drilled through these cross pieces and through the plywood so that heavy bolts can be dropped through the holes. Corresponding holes have been drilled in the top of the bench rest permitting the entire assembly to be fastened securely to the bench top.

When locked into place on the bench rest top, this assembly functions as follows: First the handgun grips are removed. The correct Ransom grip adaptors are then fitted into place in the machine rest and the handgun locked into these grips. The trigger activating lever is adjusted within the trigger guard and the firing arm positioned alongside the rest. Immediately upon firing, the handgun recoils within the machine rest. The amount of handgun recoil movement depends upon the friction fit of the slip clutch and upon the violence of the recoil. The shock of recoil goes into the plywood sheet and is then transmitted to the bolts that lock the sheet to the bench top. The heavy planks of the bench top then continue to absorb the recoil. The total effect of all this is to dissipate and absorb the shock and recoil.

The handgun is then pushed back down into battery, ready for the next shot. It is necessary to "shoot" the machine rest

Figure 6-3. Various components of the Ransom Machine Rest and a Thompson/Center pistol ready to be fitted into the machine rest.

Figure 6-4. A Smith and Wesson revolver in the Ransom Machine Rest; thumb is on trigger actuator.

into position prior to running actual tests. I usually fire a dozen rounds rather rapidly to check the point of bullet impact at the target and also to double-check functioning of the chronograph. Upon completion of this string of shots, I check the tightness of the bolts and the stability of the plywood sheet. Once I am satisfied that nothing has shifted or loosened, I commence firing for record.

There is no way that the functions of the Ransom Machine Rest can be duplicated through hand-held firing even though sand bags are used and every possible precaution is taken to insure that every shot is let off perfectly. Ammunition and the handgun in which it will be fired can only be correctly tested by using a machine rest like the Ransom. The mounting of the machine rest must be correctly done. The bench rest must be correctly designed, built, and maintained. Without all three factors being right plus the intelligent use of the machine rest, results cannot be considered valid.

THE CHRONOGRAPH

When we evaluate various types of ammunition and handguns with varying barrel lengths, it is important to know the velocities involved. A good chronograph is a must. What chronograph is purchased depends upon the amount of firing that will be done and upon the information desired. Today we have many fine chronographs and I have used several with varying degrees of success. Many are accurate but slow to use. Some use tapes, others use rays of light to measure the bullet velocity. I do not wish to become involved in the pros and cons of the many types of chronographs. At this time I am using the new Oehler Model 33 Chronotach. This is a superb instrument that provides direct velocity readings with each shot fired. In addition to providing instant direct read-out, the Model 33 gives the following information in the following order:

1. Lowest velocity
2. Highest velocity
3. Maximum velocity spread — the difference between the lowest and the highest velocities
4. Average velocity of all shots fired

5. Standard deviation

It is a small instrument, light in weight, compact, and capable of handling hundreds of rounds of recorded firings per day. It may be moved around, set up at any range to provide velocity data at the muzzle or at any desired yardage. The problem, of course, is to obtain sufficient accuracy with the weapon at hand to place the bullets within the activating area of the skyscreens.

I have worked through many setup variations and details until I now have a welded iron skyscreen holder that may be moved as required to meet the changes in handgun setup within the Ransom Machine Rest.

Figure 6-5. The chronograph and machine rest in place for testing; blast shield and steel chronograph frame extending ahead of the bench rest are also in place.

One problem with the Model 33 Oehler is that too much light will create false velocity figures, especially if there are trees or falling leaves or too much sunlight. To correct this problem, I use diapers held by the movable wood rods hanging from the tops of the two pipes to diffuse sunlight and to eliminate any accidental triggering of the skyscreens. It works and enables me to improvise at any time without worrying. I could have a complete frame built over the entire setup, but at this time I see no need for it.

Figure 6-6. The chronograph frame with its steel plates to prevent bullets from striking the Skyscreen. The sticks hung from the vertical two-inch pipes are fully adjustable; when necessary, diapers can be fastened to them to eliminate shadows and too bright sunlight on the Skyscreens.

Obviously, if only a few rounds are to be chronographed every couple of weeks, then there is no need to invest in an expensive piece of equipment. Time, the volume of firing, and other details should be worked out ahead of time prior to purchasing a chronograph; however, I sincerely believe that this Oehler Model 33 gives the shooter far more for his money than any other comparably priced machine in the business.

THE CLEANING AND MAINTENANCE OF THE TEST HANDGUNS

It was evident that there would be problems with factory

ammunition. I believe that any other make of ammunition that met the specifications of the Remington and Federal loadings would also present problems. Keep in mind that this is full-power, high-velocity ammunition. As such, it would appear that reliability and functioning would take precedence over barrel fouling and powder residue. It is vital that the shooter realize the inherent problems.

The Dan Wesson revolver that fired most of this ammunition took a terrible beating. It stood up under this beating and continued to function correctly throughout the tests. I believe that I put it through ten times as much firing as it would receive in the hands of the average shooter.

The most serious problem encountered was the barrel fouling and leading. It was obvious that I could not conduct the tests without cleaning the barrel thoroughly after each six shots. I ran a wire brush through the barrel half a dozen times after each six shots. After each twelve shots, I also ran the brush through each cylinder chamber. This did not begin to eliminate or reduce fouling and leading. It did manage to prevent the abrupt buildup that occurred before I cleaned the barrel.

The Smith and Wesson Model 13, the Ruger, and the Colt Mark III also fouled and leaded badly. I found that it was necessary on all these handguns to clean them immediately

Figure 6-7. Leading and fouling encountered in firing the four revolvers.

after firing. Then I would clean them again on each of the following four days before the barrel stopped exuding black and green fouling.

As for the actual cleaning, I used the new Hoppe's synthetic patches, Hoppe's #9 solvent, and Penguin's brass brushes. Anyone firing this type of ammunition should make certain that the handgun and the handgun barrel are thoroughly cleaned not only the day of the firing but also on each of the following two or three days. If this is not done, barrel fouling and leading will build up to destroy accuracy. In addition, all of this ammunition left considerable unburned powder and residue in the barrel, in the chambers, and throughout the area ahead of the cylinder, including the inside surfaces of the crane, the locking lugs, and the extractor rod; some was found under the extractor head. This means that, without adequate cleaning, the revolver could easily lock up or, conversely, fail to close, thus rendering the piece inoperative. Some ammunition will so coat the inside surfaces of the crane that the cylinder cannot be swung shut. This residue must be watched and tightly controlled.

The entire top of the bench rest was coated with unburned powder as were all objects on the bench rest top. Along with the terrific muzzle blast from the two- and four-inch barrels of the Dan Wesson, the unburned powder often tended to confuse the Oehler Skyscreens. I was forced to set up a blast board, consisting of a piece of plywood with a two- by four-inch hole cut into it, set up ahead of the machine rest so that the handgun barrel would put the bullet through the hole. The hole had to be relatively large vertically in order to permit the raising and lowering of the handgun muzzle to allow firing three or four groups on the paper at fifty yards. Once the blast board was put into place, most of the bad readings and problems with the Model 33 chronotach disappeared. All groups and velocities fired up to that point were then re-fired and the first results discarded.

I would like to emphasize that weapon maintenance is a very important part of firing these high velocity loadings regardless of the type, size, or quality of the piece.

Figure 6-8. Pock marks from unburned powder on the surface of the wooden blast shield.

Lead

The use of lead in handgun cartridges is so common that few shooters give it any thought. Lead is an ideal substance for handgun bullets. As a metallic element, lead is an interesting item. It is malleable and may be beaten or formed into many shapes. The Lewis and Clark expedition went into the west in the early 1800s carrying much of their reserve supply of powder in lead containers. These containers kept the powder dry and protected it from flame and other contamination. When the powder had been used up, the lead could then be melted down into ingots or bullets. Lead may be melted readily and without difficulty. It can be cast into just about any desired form. It has many desirable attributes but it does have one bad fault. When allowed to remain in contact with air and acids, it often oxidizes and finally crumbles into a dust or powder. Examine any old lead bullet and note the white powder that has formed on the outside.

Lead is the ideal bullet material (with the possible exception of gold, which would probably produce an even better bullet). The softer the lead, the more effective a bullet will be upon striking the target, allowing it to mushroom and produce maximum impact shock. Unfortunately, the softer a bullet is the more it will tend to lead the handgun barrel. As a result, a compromise must be reached between lead hardness, velocity, lubrication, and barrel smoothness.

Lead may be hardened by the addition of tin or antimony. Most lead bullets for match or conventional shooting are cast one part tin to ten parts of lead. At low to medium velocities, these bullets perform very well and generally leading is not a problem. When I was swaging match .30 caliber bullets in my Biehler and Astles dies, I bought lead wire that contained 2 percent antimony. This gave me an excellent core for the bullets that could be worked, cut, and swaged and that stood up under the .30-06 and .308 Match bullet velocities.

In 1949-1950, Jim Harvey of Lakeville Arms encased a pure lead core within a malleable gilding metal jacket to produce the first production jacketed handgun bullet. This concept was not new, having been used in the construction of rifle bullets since the turn of the century. What was new was the application of this principle to handgun bullets. The jacket permitted high velocities without lead coming into contact with the barrel, thus eliminating barrel leading. The pure lead cores expanded drastically at these high velocities, thus slamming maximum impact shock into the target. Today, such a bullet is commonplace. In the early 1950s, it was basically unknown.

Even today, I believe that more lead bullets are being fired than jacketed bullets, the latter type bullets being used by hunters and in law enforcement circles. As far as I know, all match loads fire lead bullets of a basic wadcutter design. So long as velocities remain relatively low (700 fps through 850 fps), leading will be negligible and lead deformation minor. Being malleable, lead conforms readily to barrel dimensions and allows the rifling to cut into the lead, thus rotating the bullet as it passes through the barrel. At low, match velocities, there are no problems. Problems do not arise until lubricants are insufficient; this occurs when velocities increase, and at this stage leading can build up readily within the barrel.

The higher the velocities, the hotter the gases become upon the base of the bullet. These hot gases have the capability of actually melting the lead on the bullet base, thus adding to the lead buildup in the barrel. Bullet accuracy is dependent to a great extent upon having a perfect bullet base throughout the entire travel of the bullet through the barrel. If the bullet base is not perfect, especially at the moment when the bullet leaves the handgun muzzle, then gases will blow by one side of the bullet base, tipping the bullet slightly and sending it off course.

Another source of bullet deformation is the barrel rifling. Because lead is malleable, if a bullet hits the rifling at high velocities, it will tend to skid before the barrel grooves and lands can grab it and force it to commence rotating. Usually, the shorter the barrel, the more inaccurate a lead high-velocity bullet will be because of the tendency of the bullet to skid. In a two-inch barrel, the bullet will often skid right through the barrel. In a six-inch barrel, there is sufficient barrel to finally control the skid and stabilize and rotate the bullet.

As may be seen, barrel rifling is vitally important in producing bullet accuracy. Up until the last twenty-five years, revolvers fired mostly lead bullets, therefore barrel rifling had been slanted towards handling lead bullets. Nowadays, a handgunner may fire low-velocity lead bullets in a match, high-velocity lead bullets (cast from extremely hard lead) for hunting and high-velocity jacketed bullets. Because of the tremendous variety of bullets available, barrel rifling cannot be produced solely for one type of bullet.

When Remington first brought out the .44 Magnum revolver cartridge, the big lead bullets had gas checks on the bases to aid in keeping down leading. The gas check protected the bullet base from the intense heat of the burning gases. It also, to some extent, tended to wipe out part of the barrel leading. This was a temporary arrangement because Remington shortly afterwards, along with all the other ammunition companies, brought out variations on the long-jacketed handgun bullet.

The big lead, gas check bullets are possibly the best big game hunting bullet available because they drive deep into the animal rather than expanding upon bullet impact and producing a relatively shallow wound. For this reason, gas check bullets remain in widespread use.

From time to time, we read about some well-meaning person who believes that, if sample bullets from all handguns are filed with the police, in the event of a homicide, the guilty party could be caught. So far, no one has been able to come up with a solution to filing the bullets in such a manner that finding the correct bullet would be a readily simple matter. Second, lead bullets oxidize so that, over a period of years, the rifling marks would slowly crumble away.

In addition, it is possible to take a handgun, fire both lead bullets and jacketed bullets from this handgun, and place them on file. Immediately afterwards, this same handgun can be fired very rapidly with perhaps two dozen high-velocity lead bullets and two dozen high-velocity jacketed bullets; the resulting barrel markings will be so altered that no laboratory could state conclusively that the original bullets match the final bullet fired. Every time a bullet is fired through a handgun barrel, some minor change takes place within the barrel. The hotter the barrel, the higher the velocities, and the more heat from the burning gases, the greater the changes will be within the barrel. It does not take many rounds to alter barrel characteristics.

AIR CONTAMINANTS

Handgun bullets come in three basic types — lead lubricated bullets, lead core jacketed bullets, and steel core Teflon®/nylon-coated bullets. Each type creates exhaust fumes when fired. Mixed in with the fumes are airborne lead contaminants. Over a period of time, this accumulation of fumes and residue can, under certain conditions, cause physical problems if people are exposed to these conditions on a daily basis. In other words, range officers and personnel who do a great deal of firing and instruction on indoor ranges are subject to these physical impairments.

Let's examine each type of bullet. The lead lubricated bullet is the worst offender. When the hot gases strike the exposed, lead bullet base, there is a tendency for the heat to melt a

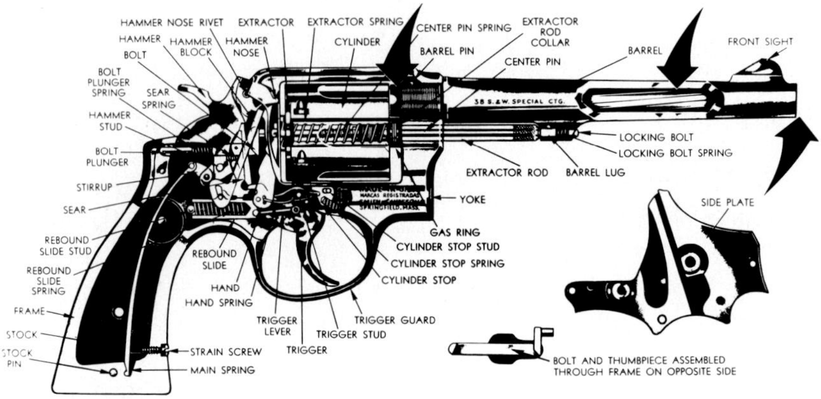

Figure 6-9. Cutaway view of the Smith and Wesson .38 Military and Police Revolver. Arrows indicate where burning gases escape from a revolver (at cylinder gap and muzzle and inside of bore accumulated lead reacts with gases of later shots). These gases carry lead and lead compounds into the air.

portion of the base, thus creating a gaseous lead contaminate. As the bullet moves down the barrel at ever-increasing velocity, the heat and friction of the barrel and its grooves will tend to produce additional lead particles that will be blown out of the muzzle along with all of the other fumes and residue. Add to this the conversion of the bullet lubricant from a solid to a hot gas or liquid and we have air contaminants that are extremely detrimental to health. This type of bullet is possibly the most dangerous type.

The second type of bullet is the lead core jacketed bullet. When this bullet is fired, the hot gases strike the base of the bullet; the jacket eliminates any creation of lead residue or fumes. As the bullet moves down the barrel, only the jacket is touching the barrel lands and grooves, again eliminating any additional buildup of lead. Since there is no bullet lubricant, these air contaminants are also eliminated. About the only source of lead in such a loading is in the primer that contains lead compounds. On the other hand, tests have been carried out that indicate that "the predominant contribution to the trapped lead comes from the projectile rather than the primer." (NBS special publication 480, *A Reduction of Airborn Lead in Indoor Firing Ranges* by A. A. Juhasz, March 1977). Another excellent dissertation on this subject is *Health Effects of Exposure to Lead in Firing Ranges* by Alf Fischbein. I also recommend *Guidelines for Police Shooting Ranges* by Ted Busch. These books and reports go into far more detail than is warranted in this discussion and I highly recommend all of them to any person responsible for indoor ranges.

The third type of bullet produces even less air contaminants, due to the synthetic material (nylon) used in coating the bullet. Again, the primer does send small amounts of lead fumes into the air, but from a practical point of view, this may be disregarded.

So far, all we have examined is the area at the firing point. Contaminants continue to move down range, forced there by the force of the muzzle blast and by the amount of firing that is going on. As a result, the entire area between the firing line and the bullet trap or backstop contains varying amounts of lead and other contaminants that have fallen to the floor or become attached to the walls and ceiling. Modern ranges are so designed that this entire area may be hosed down each morning prior to firing, thus eliminating much of the contamination that has accumulated since the last firing session. Older ranges allowed the buildup of this nonflammable residue as well as unburned powder that could accumulate until, under the right conditions, it would burn with a pale, flickering blue flame.

Despite the accumulation of lead between the firing line and the bullet backstop, most of the lead contamination occurs at the backstop. When a lead bullet strikes the steel plates of the backstop, a condition similar to the passage of the bullet through the handgun barrel is created, due to the impact shock of the bullet against the steel and the tearing and deforming of the lead by the steel plates.

We thus have three places where airborn lead contamination exists — at the firing point, at the backstop, and in the general area in between. Until recently, no in-depth study of these problems had ever been conducted; hundreds of people throughout the country have been subjected to potentially dangerous firing situations over a period of many years. Now that the dangers of lead contamination have been studied and evaluated, many ranges have been closed for alterations and for the installation of proper ventilation systems. The situation in general is far worse than expected.

An interesting side light to all this is the development by Smith and Wesson of a new type of handgun bullet, a lead bullet coated with nylon. Initial tests indicate that this ammunition is extremely free of airborne lead contamination, reducing the amount of contamination as much as 82 percent when compared to the conventional lead lubricated bullet. This, in itself, could permit the reopening of many ranges allowing the existing ventilation system to handle the new level of air pollution.

These bullets have proven to be even cleaner than lead core jacketed bullets. A combination of these new Smith and Wesson bullets plus a new, correctly designed range would just about eliminate the dangers of airborne lead contamination. This is an interesting field and one with which every range officer and instructor should be familiar.

Nyclad® by Smith and Wesson

Smith and Wesson has adapted and modified an old principle to some new ammunition. They are coating both low-velocity and high-velocity bullets with nylon, hence the name Nyclad.

It would appear that the new Smith and Wesson ammunition causes far less (61 to 89 percent) air contamination than

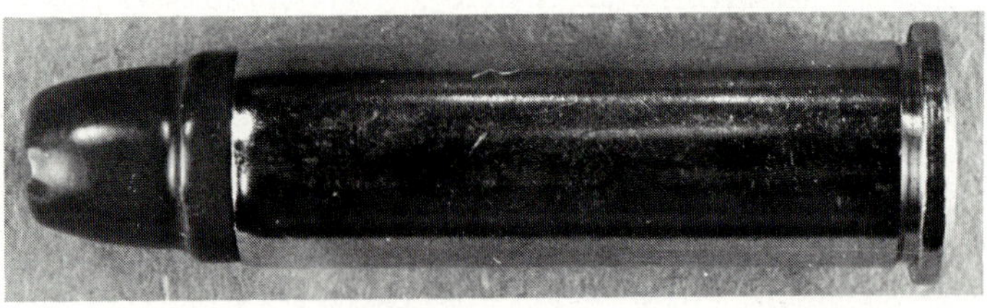

Figure 6-10. Nyclad nylon-jacketed bullet.

Figure 6-11A. Stroboscopic photograph "freezes" bullet in flight and shows smoke generated from firing conventional lead bullet ammunition. This was shot 43 of gun 2.

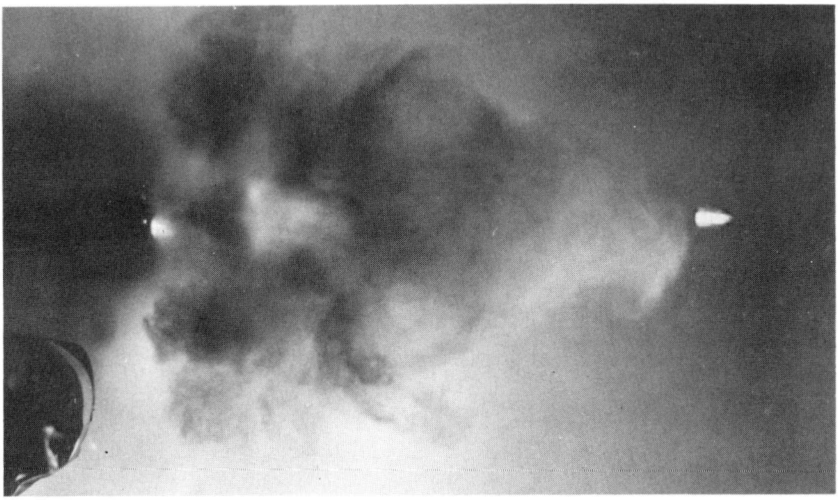

Figure 6-11B. Stroboscopic photograph "freezes" bullet in flight and shows lack of smoke when new Smith and Wesson Nyclad ammunition is fired. This was shot 43 of gun 1.

conventional lead bullet ammunition. I did not run in-depth or exhaustive tests on this ammunition. I have an indoor range and fired several rounds of Nyclad ammunition; it was obvious that air pollution was far less than when firing my conventional target handloads. By eliminating the bullet lubricant, by coating the lead bullet with nylon so that the hot gases cannot touch the bullet lead, and by preventing any lead from coming in contact with the barrel during firing, it is obvious that a substantial reduction in air contamination has been achieved through use of this nylon coating concept.

I was, however, vitally interested in side effects. First, what would nylon do to handgun barrels? I have, in other sections of this book, discussed the effect of firing mixed types of ammunition through a barrel and the resulting fouling and loss of accuracy. I wanted to determine whether or not nylon would, in any way, cause fouling or if it would strip off the bullet if the barrel were extremely hot from rapid firing or for any other cause.

We fired thirty-six rounds as rapidly as possible in a standard Model 10 Smith and Wesson with lightweight four-inch barrel. We had no problems with functioning or extraction. The barrel had been thoroughly cleaned and dried prior to firing. When the thirty-six rounds had been fired, the air in the vicinity was not polluted nor was there the customary acrid smell of lubricant with its accompaning haze. On checking the barrel, it appeared perfectly clean. A dry patch produced no residue of any kind. It would appear that there are no problems with barrel fouling when using this Nyclad ammunition.

Next, I wanted to discover what would happen if Nyclad cartridges were fired along with other types of cartridges. Would the different types, when mixed together, produce residue or fouling in excessive quantities. The following cartridges were fired:

1. Nyclad .38 Special 158 grain SWC +P
2. Super Vel 110 grain JSP
3. Federal SWHP 158 grain lead
4. Federal 125 grain JHP
5. Speer 158 grain JSP
6. Speer 125 grain JSP

7. Remington 95 grain JHP
8. Remington 158 grain Metal Point
9. Remington 200 grain lead
10. WRA 200 grain lead
11. WRA 158 grain Police lead bullet
12. WRA 150 grain Lubaloy

Firing was conducted by mixing three Nyclad cartridges with three other cartridges (of the same kind) in the cylinder. First, three Nyclad cartridges were seated in alternate chambers; then one cartridge type was seated in the other three chambers. The six rounds were then fired as rapidly as possible into wood. Wood was chosen because it has a tendency to strip jackets and deform bullets. I wanted to find out how much abuse this nylon coating would take.

One man would fire as fast as possible, dump the empties, and another man would reload the revolver cylinder. Each full cylinder contained three Nyclad cartridges. In this way, we fired seventy-two rounds within just a very few minutes. Out of this firing, thirty-six of the cartridges were Nyclad.

Having remarked upon the tendency of conventional ammunition when fired rapidly and mixed to severely foul and coat the barrel, we were surprised to note how clean the revolver barrel was when we finished firing the seventy-two rounds. A couple of passes with a wire brush, a couple of Hoppe's soaked patches, and the barrel was clean without any traces of leading or fouling.

We examined the wood blocks and the bullets in the wood blocks. Some of the Nyclad bullets had lost most of their nylon coating. Others retained quite a bit. It would appear that there were no problems within the barrel and that these bullets performed as well as any of the other bullets fired into the wood blocks.

I remain quite impressed with the barrel condition. It would appear that the nylon coating tends to wipe out barrel leading and barrel fouling. This tendency combined with the known ability of high-velocity jacketed bullets to partially clean a barrel of fouling and leading indicates that these new Smith and Wesson bullets may be the answer to many problems, including high-velocity lead bullet loadings for both law enforcement work and on the hunting fields (the characteristics of a high-velocity, hard lead bullet prove ideal for big game hunting).

I hope that Smith and Wesson will make these bullets available to handloaders before too long. It will be interesting to see what will develop in the way of factory hunting ammunition and along the lines of components for the handloader. The existing cartridges appear ideal for firing in indoor ranges. They are *not* suitable for law enforcement street work at this time: penetration remains excessive, along with a rather complete lack of bullet expansion. These problems can be corrected without too much trouble. Basically, I believe that Smith and Wesson has done an excellent job in producing these cartridges.

Part II
Exterior Ballistics

The moment the bullet leaves the handgun muzzle and enters the air, we are dealing with exterior ballistics. If the bullet leaves the muzzle with a perfect base, then it should fly accurately to the target, assuming of course perfect bullet design and construction plus sufficient rotation to stabilize it. Rotation is vital to accuracy because, as is the case with a top, when rotation slows below a certain RPM, both the top and the bullet will begin to yaw, thus allowing the top to rotate erratically and finally fall to the ground; in the case of the bullet, the bullet will fall off the line of flight and fly an uneven course until it, too, falls and strikes the ground.

Gravity affects the bullet's flight, as does wind. Usually, handgun shooting distances are so short that neither need be seriously considered. Far more important to the handgunner are the various objects that can stand between him and the target that will cause bullet deflection or even bullet stoppage before the desired target is reached.

Obviously, we must assume that bullet construction is capable of handling rotation, velocity, and bullet deformation. These problems seldom arise today but not too many years ago jacketed bullets were unable to withstand the torque or the rips in the jackets caused by barrel rifling. It took a few years for handgun bullet manufacturers to determine bullet jacket specifications that would allow the bullet to remain relatively unharmed by the time it reached open air. An example of this is to fire a .357 Magnum, jacketed, soft-point handgun bullet in a Remington .350 Magnum rifle. If full rifle loads are used, the bullet can seldom withstand the strains and stresses and will blow up or simply veer off course and disappear not too far ahead of the rifle muzzle.

In order to obtain accuracy, there must be a planned consistency in the handgun cartridges so that the bullet will be subjected to precisely the same velocity, rotation, and stresses as all of the other bullets. Changing any one cartridge component can affect bullet accuracy. It is surprising how much variation there can be in the powder charge without seriously affecting bullet point of impact. Variations of three- and four-tenths of a grain in a gross powder charge of perhaps seven grains does not necessarily mean poor accuracy. It appears that bullet perfection combined with the correct bullet design are possibly the most important factors affecting bullet accuracy. Like everything else in ballistics, this is not always true but over the years it has appeared to be the main controlling accuracy factor. The basic fact remains that if you want top accuracy, start with the bullet.

As long as nothing is permitted to touch or deflect the bullet from the time it leaves the handgun muzzle to the time it strikes the target, exterior ballistics should not be of too much importance to the handgunner. Today, most barrels are straight, smooth, and carefully controlled so that bent barrels are not a problem as is so often the case with rifle barrels. Chambering does not seriously affect handgun accuracy, although in the case of revolvers and often of pistols, custom gunsmiths completely realign and alter chambers so that they will line up perfectly and deliver the bullet down the center of the barrel, thus ensuring that the bullet leaves the handgun muzzle in as perfect condition as possible. This adds just a bit more accuracy for long range handgun hunting and competition.

Basically, exterior ballistics for handguns is far less important than either interior or terminal handgun ballistics; however, we will examine some problems that can and do arise in exterior ballistics.

Chapter 7

BULLET NOISE

WHEN a projectile leaves the handgun muzzle, there is a rush of air back into the barrel to fill the void created by the explosion and by the ejection of the bullet and the gases. There is an accompanying noise of varying intensity.

Once the bullet has cleared the handgun muzzle, it will proceed at either above or under the speed of sound. As long as the bullet velocity exceeds that of the speed of sound, there will be a distinct sharp crack created by the bullet. If the bullet velocity is below the speed of sound, this noise will not be present and the bullet will move silently to the target.

One of the reasons special ammunition must be used when using a silencer is that the purpose of the silencer is defeated if the bullet is moving faster than the speed of sound. The distinctive crack of the bullet as it moves through the air faster than the speed of sound will indicate the use of a silencer and pinpoint the location of the handgun. As a result, all ammunition for use in silenced handguns is loaded to provide a bullet velocity below the speed of sound.

Revolvers cannot be silenced for all practical purposes. In special instances, the fit of the front of the cylinder to the rear of the barrel has been hand assembled to eliminate the normal gap that exists in production revolvers. Such a modification can provide reasonable silencing for four or five shots; because of the lead and residue buildup between the cylinder and the barrel, additional firing is usually difficult or impossible because the cylinder cannot be turned. Due to these factors, revolvers are generally not used if a silenced handgun is required.

Instead, the shooter turns to automatic pistols with a locked breech for maximum efficiency. When properly assembled and fitted with an effective silencer, an automatic pistol is nearly silent when fired. One of the problems is the ejection of the empty cartridge cases; they often make far more noise when striking a wall, floor, or some other object than the muzzle report.

For this reason, single shot pistols with closed breeches are often used if a single shot will suffice. In many cases, .22 caliber bolt action rifles are cut down and fitted with a silencer and a low power scope, thus providing absolute silence plus positive sighting, in addition to allowing a succession of shots to be fired without reloading. Because of the high velocity of current .22 caliber rim fire ammunition, it is necessary to make certain that bullet velocities are below the speed of sound.

Any rifle or handgun can be suspect if the muzzle is threaded or cut off or altered in any manner. In addition, check out "barrel extensions." These are hollow tubes fitted and usually silver soldered to the end of a barrel that may be less than eighteen inches for some perfectly legitimate reason. Adding the barrel extension makes the piece legal; many times, however, the inside of this extension is threaded and altered to take a removeable silencer.

Chapter 8

VELOCITIES

REVOLVER VELOCITIES

THERE are many things we need to know about handgun velocities. The Oehler Model 33 Chronotach can supply most of this information. As a result, the following velocity figures are broken down into five classifications:

1. Minimum velocity of all shots fired
2. Maximum velocity of all shots fired
3. Extreme velocity spread of all shots fired
4. The average velocity of all shots fired
5. The standard deviation of all shots fired

In addition, I have added the group size because I believe this is a basic factor in ammunition/handgun evaluation. I have also listed the average group size of all shots fired.

When dealing with revolvers, we must remember that there are three basic barrel lengths: two, four, and six inch. Each class must be evaluated based upon its velocity average and group average.

All firing was done from a Ransom Machine Rest. The distance from the front of the bench rest to the face of the targets was fifty yards.

The complete breakdown of velocity figures and group sizes is listed below:

AMMUNITION	MINIMUM	MAXIMUM	EXTREME SPREAD	AVERAGE VELOCITY	STANDARD DEVIATION	GROUP SIZE (INCHES)
Remington, .357 Magnum — 2 inch Dan Wesson Barrel						
125 M1	1188	1252	64	1224	23	10 1/4
158 M2	1061	1136	75	1095	25	7 5/8
158 M3	1086	1124	38	1105	18	9
158 M4	784	903	119	830	43	7
158 M5	992	1069	77	1034	28	7 1/2

Average group size: 8.27 inch.

Federal, .357 Magnum — 2 inch Dan Wesson Barrel

158 A	985	1055	70	1017	27	7 1/4
125 B	1087	1157	70	1119	34	8 1/2
158 C	1065	1122	57	1087	23	6 3/4
110 D	1235	1356	121	1333	48	9

Average group size: 7.87 inch.

Remington, .357 Magnum — 4 inch Dan Wesson Barrel

125 M1	1382	1493	111	1431	38	3 3/4
158 M2	1189	1221	92	1204	11	6 1/2
158 M3	1178	1230	52	1215	20	6 1/2
158 M4	1133	1184	51	1156	20	9
158 M5	1166	1242	76	1207	26	10 1/2

Average group size: 7.25 inch.

Federal, .357 Magnum — 4 inch Dan Wesson Barrel

158 A	1146	1207	61	1183	25	6
125 B	1273	1345	72	1312	24	7 3/8
158 C	1200	1247	47	1218	19	5 1/2
110 D	1339	1371	32	1357	13	8

Average group size: 6.74 inch.

Remington, .357 Magnum — 6 inch Dan Wesson Barrel

125 M1	1462	1550	88	1505	32	6 1/2
158 M2	1270	1316	46	1300	26	2 5/8
158 M3	1229	1294	65	1261	24	4 1/2
158 M4	1234	1284	50	1256	20	5 1/2
158 M5	1213	1265	52	1234	22	4 1/2

Average group size: 4.71 inch.

Federal, .357 Magnum — 6 inch Dan Wesson Barrel

158 A	1185	1245	60	1214	19	2
125 B	1329	1383	54	1363	19	3 1/2
158 C	1223	1271	48	1248	15	4 1/8
110 D	1359	1465	106	1416	36	5 3/4

Average group size: 3.84 inch.

Just as we have definite variations in velocities and group

sizes with different length barrels, we have a substantial amount of variation between supposedly identical revolvers. In order to illustrate these differences, we took four, four-inch barrel .357 Magnum revolvers and fired twenty shots for record with each revolver. The ammunition used was the same ammunition fired in the preceding velocity tests.

Figure 8-1. Left to right: four inch barrel Colt, Ruger, and Smith and Wesson.

The revolvers used were the Dan Wesson, a Smith and Wesson, a Colt Trooper, and a Ruger. All revolvers were standard, double action, service type handguns. The results are shown below:

Remington Ammunition	*Dan Wesson*	*Smith & Wesson*	*Colt*	*Ruger*
125 M1	1387	1422	1452	1476
158 M2	1183	1238	1248	1254
158 M3	1144	1095	1230	1243
158 M4	1172	1089	1224	1223
158 M5	1174	1347	1252	1245

Federal Ammunition	*Dan Wesson*	*Smith & Wesson*	*Colt*	*Ruger*
158 A	1153	1470	1153	1232
125 B	1264	1551	1312	1361
158 C	1291	1237	1217	1271
110 D	1311	1379	1393	1413

PISTOL VELOCITIES

Pistols do not have the variety of handgun/ammunition variables found in revolvers. Ammunition can be tailor made for pistols and so extract the maximum velocity and impact shock. Pistol cartridges are high intensity cartridges loaded for known barrel lengths and for perfect functioning. As an example, barrel length variations in factory .45 ACP pistols are so minor that they can be disregarded for all practical purposes. Powder charges, bullet crimp, and powder with the correct burning rate for a pistol barrel may all be set up in advance. As a result, pistol cartridges are basically far more efficient than revolver cartridges with their inherited black powder cartridge cases.

Once again, both velocities and accuracy tests were run under the same conditions that governed the revolver ammunition tests, at a distance of fifty yards. All pistols were fired from the Ransom Machine Rest. All velocities were obtained from the Oehler Model 33 Chronotach. A minimum of ten shots were fired for each velocity figure. In some instances, groups were fired again to check results where the first series of velocity figures did not appear in line with expected results or where group sizes were questionable.

None of the pistols fired were match pistols. These pistols were fired precisely as they came from the factory.

Remington, Caliber .45 ACP — Colt Government Model Mark IV Pistol

Ammunition Reference: R45AP 1 — 185 grain wadcutter metal coated bullet
R45AP 2 — 185 grain jacketed hollow point
R45AP 3 — 230 grain metal case, target
R45AP 4 — 230 grain metal case high velocity

AMMUNITION	MINIMUM	MAXIMUM	EXTREME SPREAD	AVERAGE VELOCITY	STANDARD DEVIATION	GROUP SIZE (INCHES)
R45AP 1	743	770	27	755	9	5 3/4
R45AP 2	925	965	40	946	18	5 1/4
R45AP 3	816	843	27	830	9	4 5/8
R45AP 4	783	830	47	808	18	8 1/2

Average group size: 6.0 inch.

Remington, .9 m/m Luger — Smith and Wesson Model 39 Pistol

Ammunition Reference: R9m/m 1 — 115 grain jacketed hollow point
R9m/m 2 — 124 grain metal case

AMMUNITION	MINIMUM	MAXIMUM	EXTREME SPREAD	AVERAGE SPREAD	STANDARD DEVIATION	GROUP SIZE (INCHES)
R9m/m 1	1113	1182	69	1146	23	4 1/2
R9m/m 2	1053	1096	43	1069	15	5 1/4

Average group size: 4.88 inch.

Federal, Caliber .45 ACP — Colt Government Model Mark IV Pistol

Ammunition Reference: 45 A — 230 grain metal case
45 B — 185 grain wadcutter metal case
45 C — 185 grain jacketed hollow point

AMMUNITION	MINIMUM	MAXIMUM	EXTREME SPREAD	AVERAGE VELOCITY	STANDARD DEVIATION	GROUP SIZE (INCHES)
45 A	756	810	54	794	20	6 1/2
45 B	652	671	19	663	7	6 1/4
45 C	852	899	47	883	15	5 3/4

Average group size: 6.17 inch.

Federal, .9 m/m Luger — Smith and Wesson Model 39 Pistol

Ammunition Reference: 9-AP — 123 grain metal case
9-BP — 115 grain jacketed hollow point

AMMUNITION	MINIMUM	MAXIMUM	EXTREME SPREAD	AVERAGE VELOCITY	STANDARD DEVIATION	GROUP SIZE (INCHES)
9-AP	1078	1120	42	1089	15	5
9-BP	1085	1152	61	1129	30	1 7/8

Average group size: 3.0 inch.

Chapter 9

BULLET INTERRUPTION

QUITE often when a shot is fired at a target, the bullet is forced to pass through an object or objects before it can strike the target. These objects can vary in substance and composition. The deflection and/or deformation of the bullet can vary widely. It is impractical to attempt to cover all objects that might be found in a bullet's path; however, there are a few that should be considered.

In my opinion, the following are the objects and materials most commonly encountered: the automobile, metal sheets, brush, mobile homes, and wood.

There are few clear cut answers to any of these problems, but each will be discussed.

We must consider these problems because officers continually attempt to stop automobiles by firing at them. In other cases, metal, such as garbage cans or shower stalls is often in the way. Brush is encountered in both urban and farm areas. Few people realize that mobile homes, trailers, and campers provide little to no protection against law enforcement ammunition. As for wood, most homes and most interior partitions contain studding. If a felon takes cover behind a door or below a window or in an adjoining room, can the officer still disable him with a handgun bullet?

I cannot overemphasize the importance of reading the following sections and, even more important, actually running tests with the handgun and ammunition carried on the job. Doing it yourself is possibly the most important thing an officer can do.

WOOD

Whether we realize it or not, wood plays a very important part in our daily lives. We walk on it, live within it, handle it. There are many types of wood, oak, pine, cedar, walnut, etc.

Buildings are built of wood with the type varying across the country. It would be impractical to fire tests using twenty or thirty different types of woods for the tests. Rather, I used two-by four-inch ends, primarily pine, larch, and similar soft woods — the types commonly used in creating buildings.

The frames of old buildings in the eastern states often are constructed of twelve by twelve hand-hewn two-hundred-year-old oak beams. Even a .30-06 would not go through these beams. On the other hand, many walls consist of two by fours or two by sixes sheeted with a single layer of clapboard siding and then finished off inside with an inch or two of plaster. Firing through such a wall presents no problems. Floors were often a single layer of one or two inch pine planking; again, no penetration problems.

Other buildings used much the same framing plus a layer of one inch planking on the outside walls, then siding. Inside, there is more heavy planking plus an inch or two of plaster. This type of wall can present problems.

Brick and stone cannot be touched with a handgun so forget trying to fire through either, especially in old buildings.

In recent years, walls and roofs have consisted of two by four framing, two by sixes or two by eights for floors and rafters. Insulation poses no problem for handgun bullets even when combined with outside sheathing and dry wall on the inside. Interior walls today are mostly drywall nailed to two by fours.

The most important thing to remember is to evaluate the area in which you work and then construct walls of the same materials that make up the majority of the buildings in the area. Learn what your handgun/ammunition combination will do.

Because two by fours basically represent most of the obstacles a bullet will have to penetrate, I constructed a panel consisting of several short sections of two by fours sheeted on one side with two layers of three-quarters inch plywood to simulate a wall section. The entire panel was well nailed and tight. I stood this up against a solid wall and fired into the two by fours edgewise from a distance of ten feet. There was no question but that the bullet would penetrate the plywood sheeting, but I wanted to find out what they would do to the securely held two by fours.

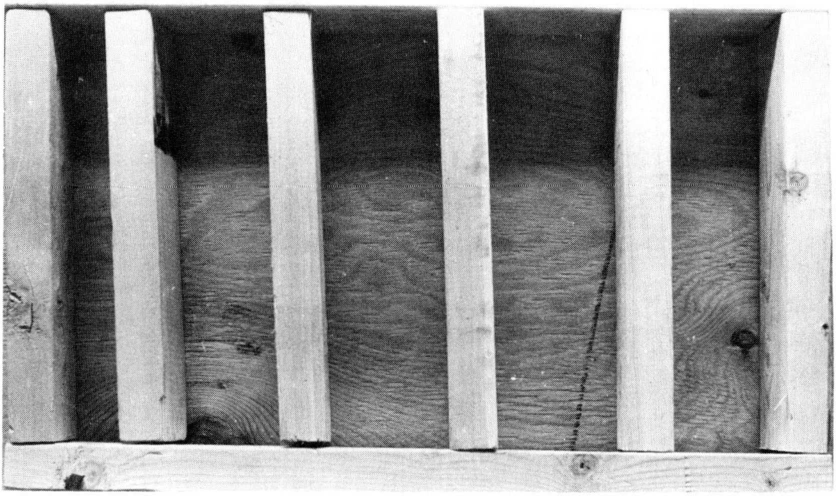

Figure 9-1. Frame work built for testing the effect of bullets against wood partitions.

Figure 9-2. The partition after testing. The .45 ACP bullets were fired into the far left upright; the bullets smashed the two by four; with the exception of a single full-jacketed bullet, they failed to penetrate. These bullets had a tendency to veer off course upon impact with wood.

The .45 bullets tended to blow the two by fours apart. The 9 mm bullets would stay inside the two by fours and some would penetrate, some would not. The .357 Magnums would generally penetrate and blow out the far side. There was no consistency in results because too much depended upon precisely where the bullet entered the two by four and how it slammed through it.

In another test, I used a section of three-quarter inch oak flooring backed by three-quarter inch plywood. All bullets passed through this.

As may be seen in Figure 9-3, bullets fired into wood usually have similar characteristics. There is no real expansion or basic deformation or lack of deformation. It would appear that bullets fired into wood can readily be identified as having

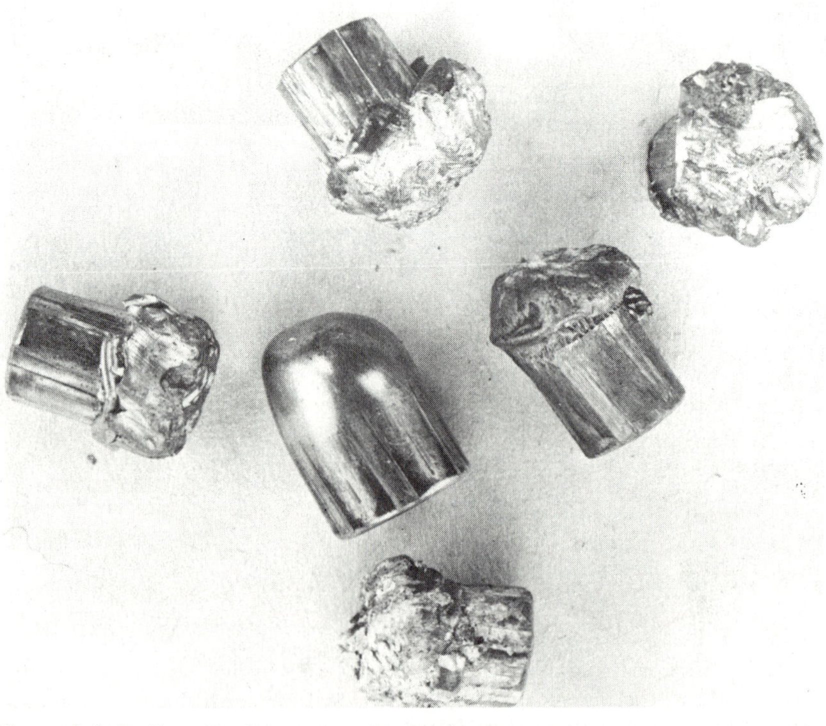

Figure 9-3. Bullets fired into wood exhibit lack of expansion and distorted noses.

passed into wood.

It would also appear that most .357 Magnum bullets will readily penetrate the walls of the average house.

MOBILE HOMES

One of the interesting post-war developments has been the mobile home, the trailer home, and the camper in all their variations. Many people regard these as a challenge to handgun bullets, placing them in the same class as the older homes. This is not true. Most are constructed of thin aluminum, two by twos, plywood, and sheet insulation, none of which offer any real obstruction to a handgun bullet.

An examination of mobile homes shows that what the housewife places on the shelves in the kitchen, the refrigerator, stove, bathtub, and other similar appliances are the only real obstacles to bullet penetration. I am sure that a .357 Magnum, 158 grain, jacketed, flat-nose, high-velocity loading would ram its bullet all the way through a fifty-foot trailer home providing, of course, it did not encounter dishes, books, or similar objects on shelves.

This is one instance where out of sight is not out of danger. A felon lurking behind a trailer wall is in as much danger as if he were standing in the open. The main problem is to keep the bullets from continuing on to injure innocent people. Therein lies a basic danger. An officer cannot fire with impunity within a trailer at a felon. He must first make certain that no one else is in the trailer in the event his shots miss the criminal. Further, he should not draw fire to himself if innocent people are behind him in other rooms of the trailer.

The flimsy construction of these structures works both ways, and considerable thought should be given to the training of law enforcement personnel who are liable to work in areas where there are trailer parks and similar structures.

AUTOMOBILES

For some reason, there appears to be a mystique regarding automobiles and the firing of projectiles into them. I believe that much of this is due to a lack of understanding of what a

.357 Magnum bullet can and cannot do to automobile bodies. Also, far too few people attempt to evaluate the potential of the bullet used in the loading. They assume that the .357 Magnum bullet can do anything. Nothing could be further from the truth, especially when dealing with cars.

Car bodies are made up of many types of metal. There are few places where a bullet can enter and continue unimpeded. There are far too many braces, struts, bolts, window mechanisms, door mechanisms, and other obstacles. Over the years, I have fired into hundreds of cars in all stages of assembly. Basically, the handgun is impotent against automobile bodies. I am speaking now only of attempting to fire through the metal or into the engine or at the gas tank. Obviously, just about any .357 Magnum will penetrate the glass windows.

I continue to go through about five cars a year demonstrating and showing the problems that exist. Once a group of officers has fired at a car, watched the results, and become involved in selective firing and long discussions, the truth of the situation sinks in. Each officer usually has his own list of myths that he must try and then try again. After continual failure, the obvious questions are asked.

I have yet to put a group through a demonstration and have anyone come up with a practical solution to the stopping of a car by firing into the metal. I am the first to admit that flukes do occur. Condemn lightweight bullets with hollow points and all of a sudden one will blast through the hood and the fire wall or slam into a door and blow out the inside fittings, but this is rare. The same applies to other loadings.

The point I am making is that ballistics for the .357 Magnum that involve automobiles can be summed up in just a few words: The handgun and its ammunition are incapable of penetrating car bodies.

An obvious question is why not use metal-penetrating bullets or even KTW ammunition. Sometimes they do work, but they are far more effective against a single sheet of metal. The intricate construction of car bodies tends to deflect the bullet nose. A bolt, a brace, anything like this, can turn the bullet point; the moment this happens, the bullet can no longer continue penetrating metal. It will spin, turn, glance, and ricochet, thus expending its energy rapidly without continuing along its original line of flight into the car interior.

As for engine blocks, I have never yet been able to put a bullet through a fender and into the engine block. I have emptied radiators, blasted fan belts, and ricocheted bullets through wiring, all of which will eventually force the car to stop. Similarly, firing at the gas tank through the bumper or trunk lid usually becomes an exercise in futility.

These comments apply to the .357 Magnum, the 9 mm and the .45 ACP. Therefore, ballistics and the automobile become informative in a negative way.

METAL SHEETS

We have discussed the effectiveness of a bullet against automobile metal. As I stated in that section, it is nearly impossible to evaluate a projectile's performance because of the lack of consistency in the obstacles a bullet will meet. In this section, I want to take up bullet performance against sheet metal only. Metal is a widely used material and the ability of handgun bullets to penetrate various types of metal can prove advantageous to law enforcement personnel.

Basically, #16 gauge sheet steel (0.0598 inch thick) is one of the most widely used sheet metals and the one I used for my tests. Obviously, regardless of what I chose, there could always be something else. This sheet metal is not comparable to special bulletproof steel any more than it is comparable to aircraft aluminum skin. On the other hand, it is a known quantity, is widely used, and should provide a working basis for the evaluation of sheet metal penetration by handgun bullets.

All firing was done from a distance of ten feet. Three sheets were used. The first sheet was held solidly against a tree by a concrete block. Six inches in front of this sheet another sheet was held tightly on another concrete block. Six inches in front of this sheet another sheet was also tightly held by a concrete block. This provided me with a large expanse of sheet metal. In addition, bullets that penetrated the first sheet would usually be found lying on the top of the second concrete block. The same would apply to the next sheet and its concrete block.

Figure 9-4. The first sheet of steel used for bullet penetration.

The penetration results are shown below.

.45 ACP

	First Sheet	*Second Sheet*	*Third Sheet*
Federal Ammunition:			
45A FMJ 230 grain bullet	Dent		
45B WC 185 grain bullet	Dent		
45C JHP 185 grain bullet	Dent		

Figure 9-5. The second sheet of steel. The bullets were unable to penetrate both sheets.

Remington Ammunition:	
R45 AP 1 185 grain bullet WC	Dent
R45 AP 2 185 grain bullet JHP	Dent (did not penetrate but did break the metal on sheet 1)
R45 AP 3 230 grain bullet MC	Dent
R45 AP 4 230 grain bullet MC	Dent

9 mm. Luger

	First Sheet	*Second Sheet*	*Third Sheet*
Federal Ammunition:			
9AP 123 grain bullet MC	Yes	Dent	
9BP 115 grain bullet JHP	Yes	Dent	
Remington Ammunition:			
R9MM 1 115 grain bullet JHP	Yes	Dent	
R9MM 2 124 grain bullet MC	Yes	Dent	

NOTE: Even though the 9 mm. bullets penetrated sheet 1, the dents on sheet 2 were very slight.

.357 Magnum with four inch barrel

	First Sheet	*Second Sheet*	*Third Sheet*
Federal Ammunition:			
357A 158 grain bullet JSP	Yes	Dent	
357B 125 grain bullet JHP	Yes	Dent	
357C Wadcutter lead bullet, 158 grain	Yes	Dent	
357D 110 grain bullet JHP	Yes	Dent	
Remington Ammunition:			
R357M1 125 grain bullet JHP	Yes	Dent	
R357M2 158 grain bullet JHP	Yes	Dent	
R357M3 158 grain bullet SPJ	Yes	Dent	
R357M4 158 grain bullet Metal Point	Yes	Dent	
R357M5 158 grain bullet Lead	Yes	Dent	

In order to determine whether two sheets placed tightly together would permit the bullet to penetrate both sheets, the following test was conducted.

Figure 9-6. A single metal sheet placed in front of a paper sheet. One shot was fired — note the impact of the lead core and the bullet jacket.

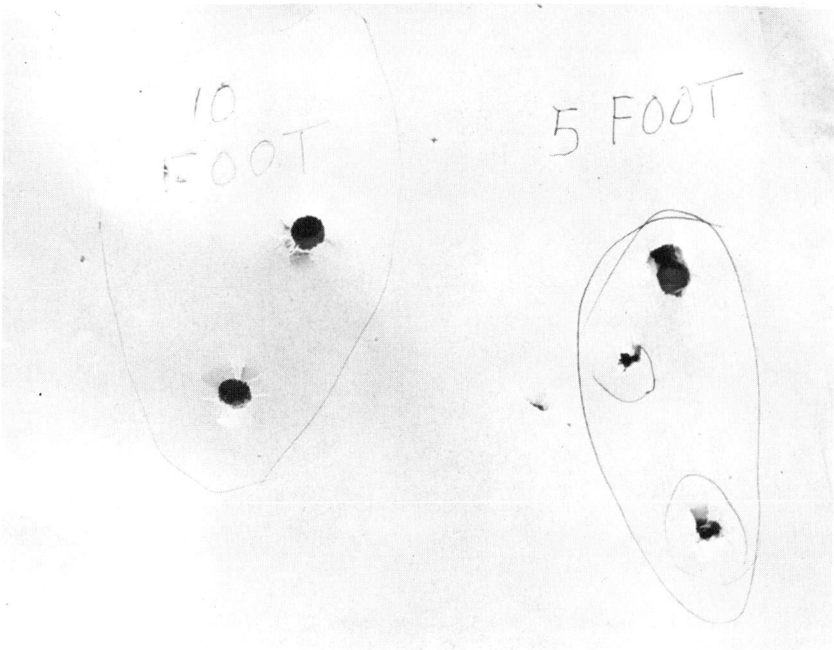

Figure 9-7. Both bullet core and jacket have a tendency to stay close to each other once the bullet has penetrated the metal sheet. The distances shown indicate the distance the sheet was from the paper; all firing was done at ten feet from the metal sheet.

Two sheets were placed tightly together, held by clamps, and placed solidly against a large tree. The Remington R357M3 bullet was unable to penetrate both sheets, giving the same results as in the preceding tests.

Next, using the Remington R357M3 cartridge, a sheet was solidly locked between two concrete blocks five feet in front of a large sheet of paper. Moving back ten feet from the face of the steel sheet, a bullet was fired through the steel sheet. Bullet particle impact upon the sheet was in a direct line with the line of fire. Following this, the steel sheet was moved back ten feet from the target face. Again, a shot was fired through the steel sheet and again the point of bullet particle impact was in line with the line of fire.

It would appear that there is little to no deviation at these distances. It also appears that all jacketed bullets shed their jackets upon penetrating steel sheets and that the jacket and the lead core both strike any object behind the steel sheet. In the above tests, as will be noted in Figures 9-8 through 9-19, both the core and the jacket continue to fly basically together.

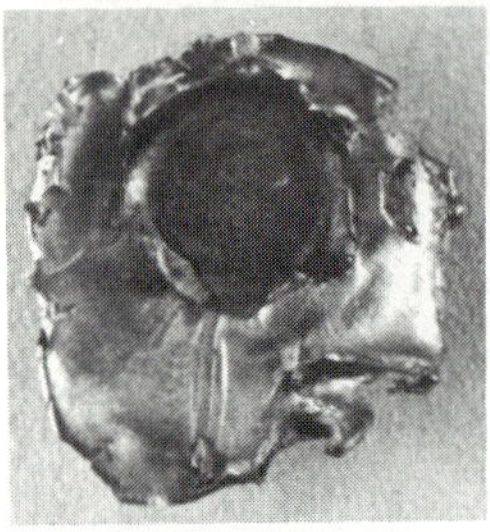

Figure 9-8. Federal .357 Magnum, high velocity, 125 grain, jacketed, hollow-point bullet fired through the metal sheets; only the jacket was found.

Figure 9-9. Federal .357 Magnum, high velocity, 158 grain, jacketed, soft-point bullet fired through the metal sheets; only the core remained.

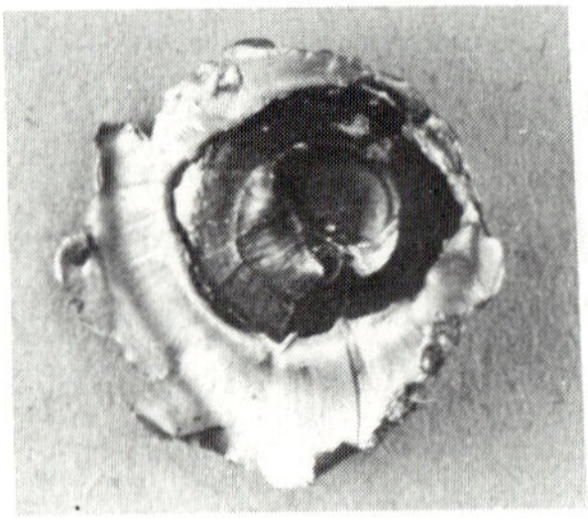

Figure 9-10. Remington .357 Magnum, 158 grain, semi-jacketed, hollow-point bullet fired through the metal sheets; only the jacket was found.

Figure 9-11. Remington .357 Magnum, 158 grain, soft-point bullet fired through the metal sheets; only the jacket was found.

Figure 9-12. Remington .357 Magnum, 158 grain, lead semi-wadcutter bullet fired through the metal sheets; the hard lead bullets without jackets remained in one piece with only partial expansion.

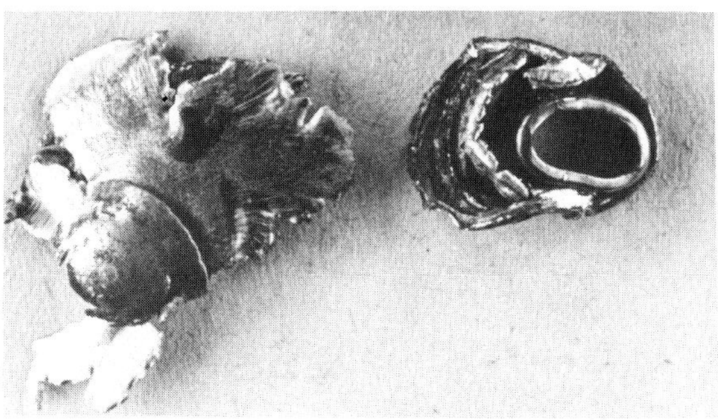

Figure 9-13. Federal 9 mm Luger, 123 grain, metal case bullet fired through the metal sheets; tremendous deformation occurs when a lightweight, full-metal-jacket bullet penetrates metal. Lead core on left, jacket on right.

Figure 9-14. Federal 9 mm Luger, 115 grain, jacketed, hollow-point bullet fired through the metal sheets.

Figure 9-15. Remington 9 mm Luger, 115 grain, jacketed, hollow-point bullet fired through the metal sheets. Completely deformed core on left; almost perfect jacket on right.

Figure 9-16. Remington 9 mm Luger, 124 grain, metal-case bullet fired through the metal sheets; the lead core fragmented completely, the jacket remained in one piece.

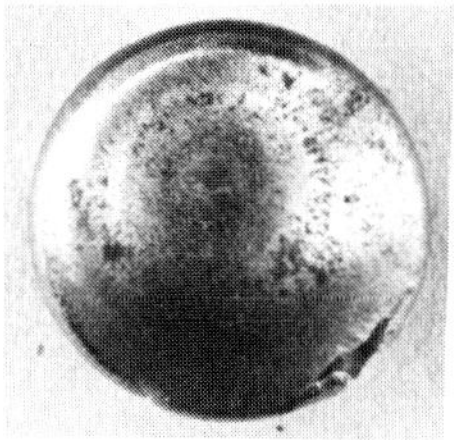

Figure 9-17. Remington .45 automatic, 230 grain, metal-case bullet fired through the metal sheets; it remained in one piece, producing a mushroom shape.

Figure 9-18. Remington .45 automatic, 185 grain, hollow-point bullet fired through the metal sheets; it mushroomed ideally and stayed in one piece.

Figure 9-19. Federal .45 automatic, match, 185 grain, jacketed, hollow-point bullet fired through the metal sheets; flattened core on the left, flat jacket on the right.

When firing through two steel sheets, note that the lead core is usually completely flattened, indicating that impact with the second steel sheet was substantial and that all stopping power and velocity have not been lost in penetrating the first steel sheet. It is interesting to note that the Remington R357M4 loading with the metal-point bullet did not produce any better penetration than the other loadings.

It would appear that any factory .357 Magnum loading will penetrate most steel sheets in the 0.060 thickness range and that special ammunition is not required.

The 9 mm Luger loadings penetrated the first steel sheet without any problems; however, their residual velocities did not permit them to do more than slightly dent the second sheet. Apparently there is not enough bullet inertia to deliver any amount of substantial shock to the second sheet.

The .45 ACP produced extremely heavy and deep dents in the first steel sheet but did not penetrate. The Remington R45AP2 firing the 185 grain JHP bullet did actually break the metal but did not penetrate. Despite the extremely large impact shock of the .45 ACP, it is not a basic metal penetrating round. As is the case with bulletproof vests, the larger the caliber, the less likely a bullet will penetrate. It is the smaller diameter, lighter weight bullets that penetrate. This must be kept in mind when evaluating bullet performance. Conversely, a high-velocity, small diameter projectile will usually have far less actual stopping power than a larger one like the .45 ACP.

BRUSH

When a projectile is in flight, its course is true. If it does not encounter any obstacles, it will eventually strike the target. If the projectile strikes a live, tough, flexible one-half inch thick limb but not nose on, i.e. the bullet ogive strikes the limb, the bullet will usually be deflected or forced to yaw, thus completely destroying its accurate flight. There are times when a bullet will strike a series of small branches and leaves and pass through them without any deflection. In other instances, a projectile will strike a piece of dry wood nose on and penetrate and still hit the target.

There are so many variations covering the ability of a bullet to strike a piece of brush that any attempt to evaluate the situation is impossible. In addition, there are so many types of brush that a bullet may encounter that, again, it is impossible to run any meaningful tests.

As an example, we can set up a series of pine rods one-half inch in diameter, forcing a bullet to pass through three feet of these rods that are set two inches apart. This is an excellent example of futility, because I have never yet found myself in a situation where I had to fire a handgun bullet through several feet of six-inch-long, half-inch-thick pine (or other type of wood) rods. Such tests bear no resemblance to firing a bullet through brush in forests.

If a situation arises on a clear, hot day in late summer that demands the firing of a revolver bullet through fifteen feet of thick yet small brush, the bullet performance and the resultant accuracy bear very little resemblance to the same situation in early spring when the same brush is whippy, growing, green, and tough. Through sheer luck, the projectile may manage to strike the target on the far side of the brush. It is more than likely that, in a similar situation, the composition of the brush will be entirely different from the brush in the first instance, thus invalidating any comparison of results.

Furthermore, if an ideal brush situation is discovered or arranged, it would be literally impossible to fire a series of bullets through identical brush situations. Each bullet would slightly alter the brush for the next projectile, thus automatically varying the results.

Brush will basically deflect a bullet from its true course. The amount of deflection will vary from shot to shot and from brush to brush. It will also vary from bullet to bullet, depending upon velocities and bullet construction. If a shot must be taken under these conditions, any bullet impact at the target must be considered as chance. Conversely, if a shot must be fired through brush with conditions requiring pinpoint accuracy, it should not be taken. If there is a hostage situation and it is believed that a shot must be taken to rescue the hostage, I would strongly suggest bringing in a high power rifle firing a heavy bullet. The odds of placing a handgun bullet into the felon and not striking the hostage are very low. Even with the rifle, they are not excellent. Bullet deflection could result in having the hostage killed.

Chapter 10

HOLLOW BASE BULLETS VERSUS BEVEL BASE BULLETS

EVERY target shooter has, at some time, wanted to know what his pet load would do at fifty yards. The day is long gone when a match shooter can sluff through with a fair fifty-yard score and then win the match with superb twenty-five-yard timed and rapid fire shooting. The key to breaking 2600 is a good fifty-yard target.

Back in 1930s and 1940s, I accepted or dropped a specific hand load based upon its ability to give me good or fair groups over a period of time. Today, with the cost of components continuing to rise, most target shooters cannot afford the added expense of running accuracy tests over a period of three or four weeks.

To further complicate the situation, we now have flat base match bullets and hollow base match bullets. The choice of bullets is not that simple because of the tremendous variety of cast/swaged match bullets on the market today. During most of my match shooting career, I used 2.8 or 3.0 grains of Bullseye and a flat base, solid base, cast bullet. This combination had been well received and accepted.

When the Smith and Wesson Model 52 match .38 Special pistol first appeared on the market, I obtained several boxes of match ammunition for testing the pistol. The results were extremely discouraging, as I had "flyers" and "tumblers." Finally, I pulled a cartridge and discovered that the bullet had a very deep hollow in the base. This loading was on the high side because it had to both function reliably in the pistol and provide accuracy at fifty yards. I pulled the bullets from a couple of dozen cartridges and, discarding the hollow base, replaced these bullets with my cast, solid base bullets. I had no more flyers

---

Adapted from *Guns & Ammo*, December 1977. Courtesy of Peterson Publishing Company, Los Angeles, California.

and obtained quite respectable accuracy at fifty yards. It became obvious that at the higher velocities the deep, thin skirts of the hollow base would often fly off the bullet base due to the centrifugal force.

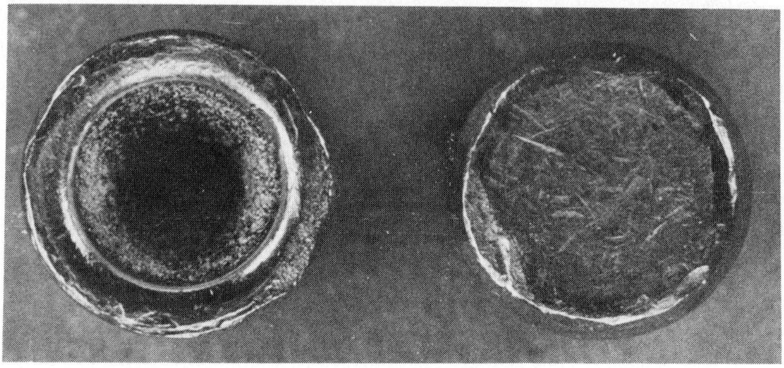

Figure 10-1.

For many years I viewed hollow-base bullets with suspicion and stayed with my flat-base wadcutters for match shooting. At that time, we had no method of determining precisely what a handgun/ammunition combination would do. True, the factories had facilities, but individuals could not afford the equipment.

In the intervening years, match bullets have changed and we had many machine rests for handguns come and go. The Ransom Machine Rest has remained as the proven and most dependable rest on the market. It is expensive, but if we want facts and not opinions, we must use the best or our tests are merely an interesting series of firings.

I have long wondered how far bullet manufacturers have come with their flat-base and hollow-base bullets, and also precisely what relation there is between the accuracy of a fine revolver and a fine match pistol. Because we now have the equipment to run meaningful tests, I decided to discover how much progress has been made in bullets and attempt to answer many other questions.

Remington sent me several hundred primed, empty, nickel cases and several thousand new, small pistol primers. Hercules furnished several pounds of Bullseye, the classic match powder. Hornady sent me several thousand bullets, half bevel base and half hollow base.

Figure 10-2. (Left) the Hornady hollow-base bullet and (right) the Hornady flat/bevel-base bullet. Both bullets are coated with a waxlike lubricant.

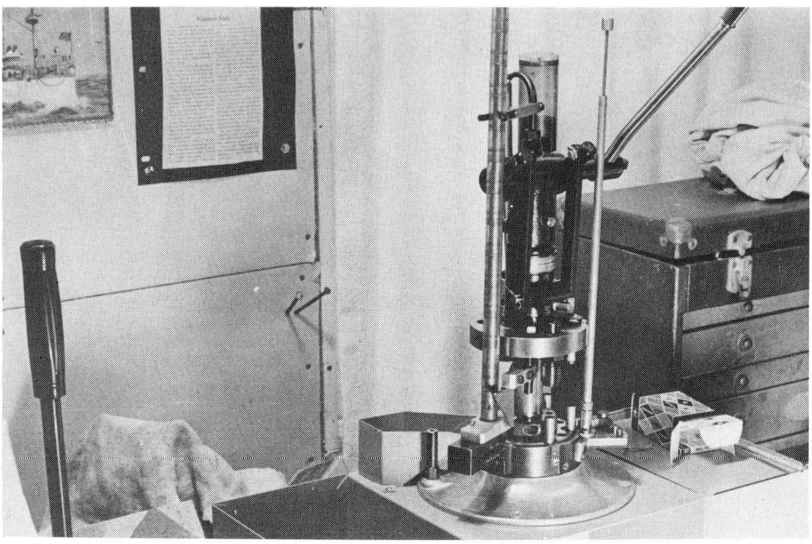

Figure 10-3. The Star reloader mounted on the special B&G Products base.

In view of the number of rounds that would be fired in the proposed tests, I realized it would be faster, and also more consistent, if the ammunition were assembled on my Star reloader. I obtained from Star Machine Works the following charge slides: 2.0, 2.2, 2.4, 2.6, 2.8, 3.0, 3.2, 3.4, 3.6, 3.8 and 4.0 grains, all for Bullseye powder.

I discussed the proposed tests with Smith and Wesson and they shipped a K-38 Match target revolver to me along with a Model 52 .38 Special pistol. I had originally considered using custom handguns, specially assembled for the tests and specially tuned, but these handguns are expensive and difficult to obtain.

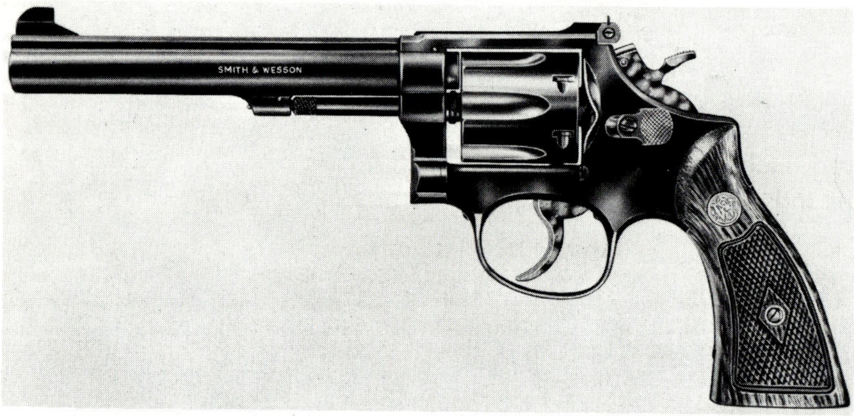

Figure 10-4. Smith and Wesson K-38 Heavy Masterpiece.

Note that all components and handguns are the best commercially available today. Anyone can purchase them. There is nothing that I used that cannot be purchased over the counter. I believe that this adds substantially to the validity of the tests.

There were many things I wanted to find out: Would a revolver prove to be as accurate as a pistol? Which bullet, the bevel base or the hollow base, would be most consistently accurate? At what velocity level would leading occur? What velocities would I be working with between 2.0 and 4.0 grains of

Bullseye and the 148 grain swaged bullet? Note that the Hornady bullets were swaged, not cast. They were not lubricated with grease but were covered with Hornady's dry lubricant that did not dirty the fingers nor build up on the seating and crimping die.

Velocities would be run on my Oehler chronograph. Because of my experience years ago with hollow-base bullets, I was curious to determine just how much practical velocity these bullets would take before the skirts let go, with resulting flyers.

I assumed that somewhere within the loading levels I had set each type of bullet would turn in excellent accuracy. Basic trends should develop that would set a pattern for loading for both types of bullets as well as for both types of handguns.

Loading procedures were kept simple. As each powder slide was changed, fifteen charges were dropped before a loading run was made. This insured that powder variations were held to a minimum during the start of a new run. I loaded sixty rounds for each powder charge for record, figured as follows: ten rounds for record with the revolver; ten rounds for record with

Figure 10-5. Cocking the revolver and preparing to fire. The lever on the right side of the Ransom Rest is the trigger lever.

Figure 10-6. The Model 52 .38 Special pistol has just been fired; the slide remains open.

the pistol; five rounds for record for chronographing with each handgun, totalling thirty rounds with each bullet or sixty rounds grand total. Backing this up were test loads for getting on paper and then a minimum of twenty rounds for shooting in the handgun prior to commencing firing for record.

We did not clean the handguns during the tests until leading became obvious and accuracy commenced falling off. This occurred at about 3.2 grains. Above that, leading was extremely bad, so bad in fact that we did not attempt to fire 3.8 or 4.0 grain loads. We did fire some for test but not for record. These high-velocity loadings are impractical for match work.

The following results are self-explanatory:

Powder Charge	Hollow Base	Velocity FPS	Bevel Base	Velocity FPS
K-38 Revolver				
2.2	2 3/8	664	4 3/4	642
2.4	3 1/4	714	5	683
2.6	4 1/4	761	5 1/2	715

Powder Charge	Hollow Base	Velocity FPS	Bevel Base	Velocity FPS
2.8	3	776	6 1/8	773
3.0	3 1/2	829	4 3/8	814
3.2	4 5/8	860	4 1/2	827
3.4	4 5/8	849	4 1/4	852
3.6	4 1/2	934	4	868
Model 52 Pistol				
2.2	3 7/8	587	6 7/8	627
2.4	4 1/2	595	7 1/8	667
2.6	4 7/8	653	6 3/8	707
2.8	3 1/2	726	7 1/4	686
3.0	5 1/2	770	6 7/8	704
3.2	4 1/4	723	6 3/8	715
3.4	6 3/4	786	8 3/8	772
3.6	4 1/4	887	4 1/4	897

I was definitely surprised to note that the revolver consistently outshot the pistol. I had always assumed that a single cartridge, lying within a tightly machined chamber and firing directly into a barrel, would prove to be far more consistent than a mechanism that attempted to revolve six chambers into alignment with a single barrel throat. I do not wish yet to retreat from my basic thinking. However, I must admit that with a moving barrel, such as is found on the Model 52, there must be a relationship between the barrel movement and the larger groups fired by the pistol. If this barrel were stationary, as is the case with the revolver, then I believe that the pistol could produce smaller groups. I plan to explore this potential later.

During most of my match shooting years, I used a solid, flat-base bullet. I won my matches with these bullets. I now wonder if I would not have won far more matches if I had used a hollow-base bullet. Many times I have called a shot good and had it go out. This proved to be true of most bevel-base bullets fired in these tests. Somewhere during the firing of the ten shot groups, one or two shots would go out of the group. It is obvious that the hollow-base bullet is more accurate than the bevel/solid-base bullet.

It is interesting to note that all velocities of the two types of bullets, when fired in the K-38 revolver, present an almost calculable escalation of velocity. This is not true of the pistol. Pistol velocities vary substantially and rise and fall despite the steady 0.2 increase in powder charges. True, the bevel-base bullet produces less space within the cartridge case for the powder; however, this space remains constant, altered only by the slightly increasing powder charge. I would have assumed that, as a result, velocities with the bevel-base bullet would have been higher due to the smaller amount of room for the powder charge within the case.

We obtained no leading with the loads using 3.0 grains or less of Bullseye. This adds additional emphasis to the importance of loads in the 2.2 to 2.6 grain area. There was no leading and extremely good accuracy. It is worth noting that the 2.8 grain loading in hollow-base proved quite accurate in both the revolver and the pistol. Loads above 3.2 grains tended to lead and accuracy fell off badly.

We did not fire the 3.8 or 4.0 loads for record because leading was so bad that the results would have proven useless. Similarly, the 2.0 grain load would not function in the Model 52 and accuracy in the revolver fell off.

The pistol functioned well throughout the tests without malfunctions of any kind. It would appear that with the specific pistol we had a 2.2 grain load is as low as it was possible to go and have reliable functioning.

The results would indicate that my thinking over the years, and the thinking of a great many match shooters, has been wrong. I have always loaded in the 2.8 to 3.2 grain area, but this is not the most accurate area.

In order to compete today, a match shooter must have a handgun/ammunition combination that will consistently place all shots inside the ten ring of the target. A handgun and its ammunition that cannot do this will cost the shooter both X's and points. In my opinion, any serious handgun shooter must either have his own Ransom Machine Rest or have access to one so that he can develop ammunition for his handgun that will do the job and produce both reliable functioning and the necessary accuracy.

So many questions have been raised that I went back over the entire program. First, I contacted Joyce Hornady who supplied the bullets. He told me, "Like yourself, we've been using the Ransom Rest for our accuracy tests with pistols, and shooting our bullets, we get our best accuracy in the revolver with hollow-base wadcutters. I am at a loss to explain the reason, but we do not expect to be able to make a bevel-base bullet that will ever shoot. We've tried turning the bevel on with a lathe so we had them as near perfect as possible, but that didn't help. We've made the bullets longer and heavier and shorter and that didn't help either. They just don't seem to want to shoot with a bevel base."

With this confirmation of my tests, I then turned to Smith and Wesson. I returned the Model 52 pistol and asked them to thoroughly check it out, and a summary of my tests along with a copy of the results. Bill Clede called me to say that he had spent some time in the plant watching the actual firing of my Model 52 in their machine rests. He sent me the actual targets with data.

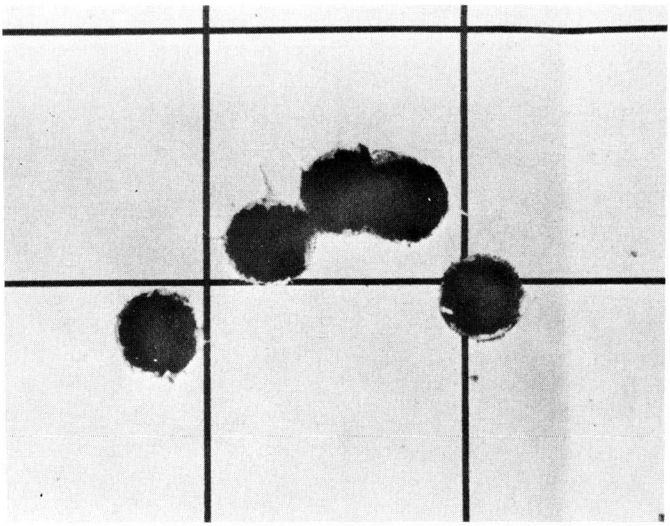

Figure 10-7. The fifth group fired by Smith and Wesson, Model 52 pistol, magazine #2. Size of the five-shot group is one and one-half inch.

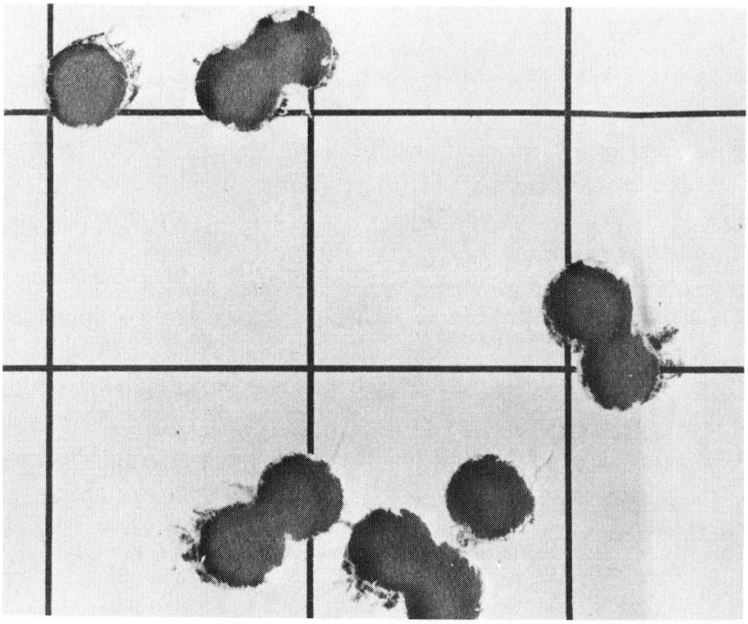

Figure 10-8. The sixth group fired by Smith and Wesson, Model 52 pistol, magazines #1 and #2. Size of the ten-shot group is two and three-eights inch.

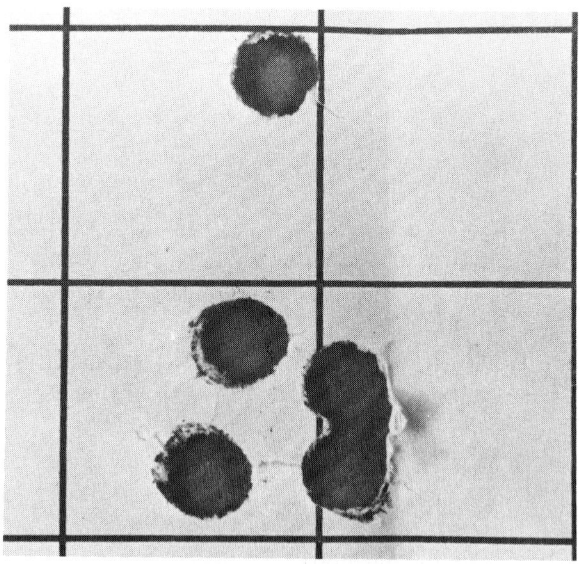

Figure 10-9. The seventh group fired by Smith and Wesson, Model 52 pistol and magazine #1. Size of the five-shot group is one and five-eights inch.

All ammunition fired was .38 wadcutter The handgun was held in a specially designed and constructed machine rest. Distance was 50 yards. Two different magazines were used. Many shooters do not realize that magazines can, and do, affect accuracy. With one make of ammunition, one magazine may shoot more accurately; with another make of ammunition, a second magazine may produce smaller groups.

Target	*Ammunition Brand*	*Magazine*	*Group Size*	*No. Shots in Group*
#1	Win W38 SIP	One	3 1/4"	Five
#2	Win W38SIP	Two	2"	Five
#3	Rem#6138	Two	3 3/4"	Ten
#4	Rem #6138	One	2 1/8"	Five
#5	Fed #38-A	Two	1 1/2"	Five
#6	Fed #38-A	One & Two	2 3/8"	Ten
#7	Fed #38-A	One	1 5/8"	Five

Note that the pistol continues to be selective in regards to both magazine and ammunition. Targets #5, #6, and #7 indicate that Federal #38-A wadcutter ammunition works best in this pistol. These groups are excellent and meet all match requirements. Target #7 with its ten-shot group fired with the two magazines compares favorably with the best groups previously listed fired with the K-38 revolver.

It would appear that the match shooter must not only evaluate various bullet makes, types, shapes, and seating depth, but he must also explore the potentials of many different powder charges and even, if necessary, different types of powders. If he is not a handloader, then he must, without fail, evaluate ammunition from all the manufacturers. It is not enough today to purchase "match" ammunition and then go out and fire in a match. The handgun/ammunition combination is just as important as the practice and physical and mental training that the shooter must undergo to become a top ranking handgun shooter.

Chapter 11

THE EFFECT OF REVOLVER CHAMBER LENGTH UPON ACCURACY AND VELOCITY

As precision handgun shooters know, revolvers fire low-velocity, match ammunition exceptionally well, despite the wadcutter cartridge with its flat-nose bullet not filling the revolver chamber. In the .357 Magnum revolver there is a substantial gap between the front end of the .38 Special cartridge and the front end of the .357 Magnum chamber. It would appear that as long as low-velocity ammunition is fired, the distance that the bullet travels from the case to the barrel throat does not affect accuracy. I have run many tests over the years and have fired hundreds of thousands of rounds in both competition and training and I believe that this conclusion is valid.

On the other hand, once velocities increase and high-velocity, jacketed bullets are used, there appears to be a falling off of accuracy and a lack of consistency in velocities. There are many people who use .357 Magnum revolvers but carry .38 Special ammunition in them. I have long questioned the validity of this thinking, so I ran a series of tests to determine if my theories are true.

Using the Dan Wesson revolver, fitted with the four-inch barrel, I set up both my machine rest and my chronograph. Ammunition turned out to be a problem because I wanted to use new factory Remington and Federal .38 Special cartridges. I did find sufficient .38 Special ammunition to run the tests using 125 grain, hollow-point jacketed cartridges.

All firing was done at fifty yards. Two six-shot groups were fired and the two first shots of each group were eliminated from both velocity figures and from group sizes because the barrel was cleaned between each six shots and it appears that cleaning the barrel affects both velocity and accuracy. Therefore, velocities are based upon two four-shot runs and eight shots on paper for group size.

In order to present the entire picture, I am listing below all data on velocities because these figures indicate that a definite trend could set in.

Ammunition: Remington 125 grain HP jacketed

.38 Special #2038 (8 shot group size: 4 5/8 inch)	*.357 Magnum ML* (8 shot group size: 4 1/8 inch)
Group 1	
Low — 1002 fps	Low — 1370 fps
High — 1136 fps	High — 1457 fps
Extreme Spread — 134 fps	Extreme Spread — 87 fps
Average — 1091 fps	Average — 1406 fps
Std. Deviation — 47	Std. Deviation — 29
Group 2	
Low — 999 fps	Low — 1353 fps
High — 1127 fps	High — 1398 fps
Extreme Spread — 128 fps	Extreme Spread — 45 fps
Average — 1075 fps	Average — 1374 fps
Std. Deviation — 48	Std. Deviation — 19

Ammunition: Federal 125 grain HP jacketed

.38 Special #38 F (8 shot group size: 3 7/8 inch)	*.357 Magnum 357 B* (8 shot group size; 3 1/4 inch)
Group 1	
Low — 873 fps	Low — 1258 fps
High — 972 fps	High — 1300 fps
Extreme Spread — 99 fps	Extreme Spread — 41 fps
Average — 940 fps	Average — 1279 fps
Std. Deviation — 38	Std. Deviation — 17
Group 2	
Low — 939 fps	Low — 1225 fps
High — 992 fps	High — 1323 fps
Extreme Spread — 58 fps	Extreme Spread — 98 fps
Average — 958 fps	Average — 1268 fps
Std. Deviation — 20	Std. Deviation — 32

When firing high-velocity loads, the chamber should be no longer than the cartridge. There is a definite lack of consistency in the .38 Special velocities that do not show up in the .357 Magnum. Both the extreme spread figures and the standard deviation figures point this up.

These tests were run quite some time after the basic velocity tests had been run. During this time, we discovered that by cleaning the barrel between every six rounds accuracy was definitely improved. The original figures will stand because they represent a practical accuracy standard, i.e. an accuracy standard that could be expected on the street and in the field. In my opinion, a great deal of work must be done with both ammunition and barrels to eliminate the fouling and leading problems inherent in current barrels and ammunition.

Chapter 12

THE RELATION OF BARREL LENGTH TO CARTRIDGE POTENTIAL

As shooters know, each factory cartridge is designed to do a specific job. As an example, the wadcutter, match loading is assembled to provide maximum accuracy at fifty yards combined with low velocities and light recoil. At the other end of the line is the 110 grain, high-velocity, +P cartridge that provides maximum velocity, maximum bullet expansion, and flat shooting at close distances. Each of these loadings plus all of the other loadings in that specific caliber differ from each other because no one loading can handle all situations and requirements.

To further confuse the situation, we have revolvers with many varying barrel lengths so that combining various barrel lengths with cartridge loadings can produce tremendously varying results. Accuracy can fall off. Velocities can increase or decrease. Powder may or may not be completely burned within the barrel. Bullet performance can be drastically altered.

In order to evaluate the basic situation, I ran tests using the following Federal .38 Special ammunition:

1. Index #38A — 148 grain, wadcutter
2. Index #38B — 158 grain, lead, round-nose standard velocity
3. Index #38D — 158 grain, round-nose, high-velocity
4. Index #38E — 125 grain, jacketed, hollow-point +P
5. Index #38F — 110 grain, jacketed, hollow-point +P
6. Index #38G — 158 grain, lead SWC HP +P

The above ammunition was new factory ammunition.

In any test or evaluation it is desirable to eliminate as many

Adapted from *Guns & Ammo*. Courtesy of Peterson Publishing Company, Los Angeles, California.

variables as possible so that the results will carry more validity. Because of this, I did not wish to use a revolver for these tests. Revolvers have cylinders containing six chambers and despite quality controls, each of these chambers is slightly different from each of the others. In addition, each chamber lines up slightly different with the forcing cone than any of the other chambers, resulting in variables that should be discarded prior to testing.

As a result, I secured a Thompson Center ten-inch barrel contender. This gave me a single, sealed breech mechanism that would allow me to run the tests with a minimum of variables. By using the same barrel throughout, I would have a constant. The same applies to using the same chamber throughout. Thus, the tests would have far more credibility than if I had used four revolvers with four different barrel lengths or even the Dan Wesson revolver.

Over the years, I have run similar tests with bolt action rifles and have watched velocities increase as barrel length is cut and then finally as the barrel length becomes too short for correct powder burning, velocities fall off. I had no idea whether the same conditions would apply to handguns. The average shooter does not realize that powder burning characteristics combined with barrel length determine velocities. If a fast-burning powder is used, maximum velocities may be obtained in a relatively short barrel. Conversely, if a slow-burning powder is used, barrel length must be increased to provide a burning chamber for the powder. Otherwise, a substantial amount of unburned powder can be blown out the handgun muzzle and also left within the barrel itself. The question naturally arises as to why various powders must be used. If a light-weight bullet is loaded into a cartridge, normally a relatively fast-burning powder will provide maximum velocity. If a heavy bullet is used, then a much slower burning powder must be loaded into the cartridge in order to hold down pressures and still give the bullet acceptable velocities. If velocities and pressures were not important, we could use a single powder for all weight bullets. Empirically, this cannot be done, so the practical solution is to select the correct combination of powder and bullet to obtain the desired bullet velocity and bullet expansion, plus safe working pressures.

I decided to run tests on velocity and on accuracy at fifty yards, because it would be possible to do so simultaneously. The Ransom Machine rest would hold the Contender pistol. We could remove the barrel without taking the pistol out of the machine rest so that the barrel could be cut two and one-half inches without any problems. The same bullet that gave us velocity figures would also give us groups on target at fifty yards. My Oehler chronograph provides me with instant readout plus other automatic calculations that are not important in these tests. I wanted to know the maximum velocity variations, velocity, and group size.

This information could be of definite importance in choosing a loading for use in handguns, particularly if several barrel lengths would be used. A two and one-half inch barrel revolver should fire ammunition basically designed for use in such a handgun. Another handgun with a six-inch barrel should fire other types of loadings that produce efficient results with that particular barrel length.

In setting up for these tests, we followed a definite routine. Before firing for record, the Thompson Contender was shot into the machine rest with at least fifteen rounds. Point of bullet impact on the target sheets was checked and corrected if necessary. We also double-checked velocities and light conditions. Even though we used light filters over the Oehler Sky Screens, we wanted to be certain that sun and shade would not affect results.

As a result of having run extensive tests many years ago with the cutting of rifle barrels, I expected that somewhere along the line some of the loadings would show an increase in velocity as the barrel was cut. This would indicate that powder burning characteristics were too slow for some barrel lengths. I also anticipated a steady decrease in velocities with most loadings. As for accuracy, I had no idea what would happen. If velocities became erratic combined with ever-increasing velocity variations, then accuracy should fall off.

We locked the Contender into the machine rest, set up the

Figure 12-1. Preparing the Thompson/Center pistol for insertion in the grips and the Ransom Machine Rest.

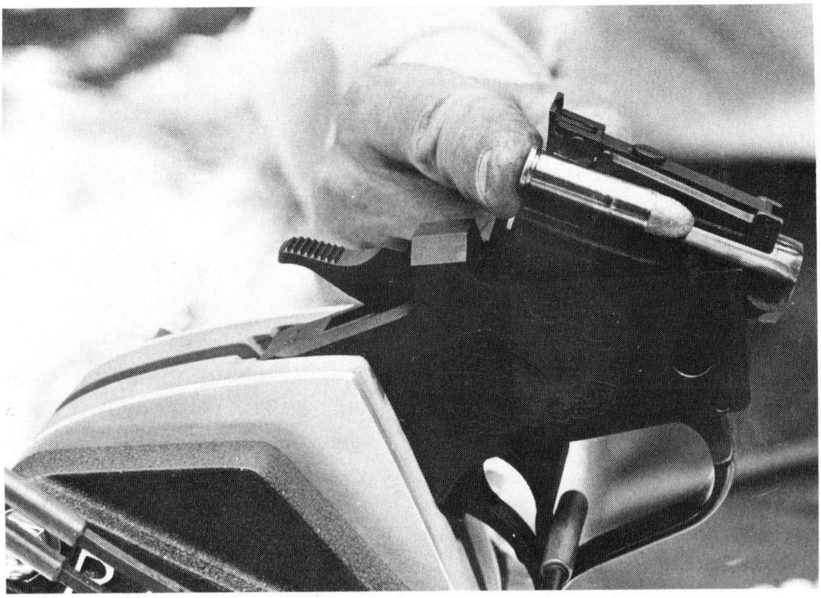

Figure 12-2. With the cartridge in place, little barrel remains.

Figure 12-3. The barrel during the final stage of firing.

Figure 12-4. The ammunition, machine rest, Redfield scope, and Freeland scope stand.

chronograph and commenced firing.

The results follow:

	38A	*38B*	*38D*	*38E*	*38F*	*38G*
Barrel Length 10″						
Velocity Variation	15	5	14	3	40	5
Velocity	834	935	1072	1229	1276	1116
Group Size (50 yds)	2″	1 1/2″	1 3/4″	3 1/4″	4″	4″
Barrel Length 7 1/2″						
Velocity Variation	21	15	30	10	45	16
Velocity	798	894	998	1140	1207	1080
Group Size (50 yds)	2 3/4″	3 7/8″	2 3/4″	2 1/4″	3 7/8″	3 1/4″
Barrel Length 5″						
Velocity Variation	3	13	4	25	18	38
Velocity	737	813	900	1035	1073	936
Group Size (50 yds)	2 5/8″	4″	4 1/4″	5″	4 3/4″	8 1/4″
Barrel Length 2 1-2″						
Velocity Variation	7	15	12	8	19	30
Velocity	627	784	694	772	819	766
Group Size (50 yds)	18″	6″	21″	5″	5 3/4″	14″

Velocity variations within a single barrel remained well within normal expectations. There was no wild fluctuation of velocities, such as 200 to 300 fps, as might have been expected, particularly with the two and one-half inch barrel. Everything considered, velocity variations remained small.

Velocities fell as expected with each barrel cutting. This did not produce any surprises. On the other hand, the accuracy figures are interesting. It would appear that the ten-inch barrel handled all the loadings better than any of the other barrel lengths.

Load 38A, 148 grain wadcutter was the most consistently accurate load in the three longer barrel lengths. It would appear that in the two and one-half inch barrel, the rifling could not grab the lead correctly and that the bullet was rammed down through the barrel, deformed and unable to rotate correctly.

Load 38B, 158 grain round-nose lead gave reasonably good

accuracy with all barrel lengths including the two and one-half inch barrel.

Load 38G, 158 grain HV +P lead SWC bullet lost accuracy rapidly in the five-inch barrel and went completely sour in the two and one-half inch barrel.

It is interesting to note that load 38A lost 207 fps, load 38B lost 151 fps, and load 38G lost 350 fps; these are all lead bullet loadings. It would appear that a moderate lead bullet loading using a heavy bullet similar to load 38A provides far more consistent and efficient functioning than either a match wadcutter loading or a high-velocity +P loading like the 38G load. This latter loading appears to be using a slow-burning powder that functions best in long barrels and which needs long barrels to produce acceptable accuracy.

Note that load 38D, also a 158 grain bullet (high-velocity loading but with a jacketed bullet), failed completely in the two and one-half inch barrel. Velocity loss was 378 fps. It would appear that this bullet blew right on through the barrel and printed on paper at fifty yards in all positions. A smooth bore barrel could not have given worse results on target. Thus, bullet construction and design become a very important part of accuracy and such a loading should not be used in short barrel revolvers.

Load 38E with its 125 grain jacketed, high-velocity hollow-point bullet and +P loading lost 457 fps. For some reason, accuracy remained acceptable in all barrels. It could well be that the light-weight, jacketed bullets are caught by the rifling of the two and one-half inch barrel and, due to the use of fast-burning powders, not only engage the rifling but obtain sufficient rotation to produce accuracy in short barrels.

Load 38F firing a 110 grain, jacketed, high-velocity, hollow-point bullet and +P loading lost 457 fps. Like load 38E, accuracy remained acceptable possibly for the same reasons. Both these loadings have been designed for use in short barrel revolvers and the components used appear to be very compatible with small size, short barrel handguns.

It would appear that the jacketed bullets of 125 grain and 110 grain are maleable enough to allow the bases to expand under pressure to allow the rifling to rotate them. Skidding appears to be held to a minimum. Obviously, rifling design has much to do with bullet rotation and the results obtained in these tests may not be applicable to other types of barrel rifling.

Unburned powder covered the bench rest top ahead of and to the side of the Contender. There was a substantial amount on the ground ahead of the bench rest.

Part III
Terminal Ballistics

We have, so far, learned how the projectile is launched from the handgun and how it flies through the air. Terminal ballistics deals with what happens when the projectile comes to a stop. This covers a wide spectrum of results. Three factors usually control what the bullet does upon finally coming to a stop: bullet construction; bullet velocity; and the composition, or make-up, of the material against which the bullet finally comes to rest.

If an officer fires at a felon at a distance of fifteen feet and if the bullet strikes precisely where the officer aims, the bullet (if correct ammunition is used) should come to a complete stop within the felon and put him down, expending all its energy within the target. However, suppose the officer fires his first shot and misses the felon so that the bullet travels past the felon to strike the hood of an automobile, ricochets off that into a window, passes through the window to enter the ceiling of a room to finally stop, embedded within a two by six rafter. Perhaps the officer, as he draws his handgun, accidentally fires a shot into the air; the bullet will move upward until gravity slows it down to bring it back to earth on a curve. When this bullet falls back, it can only travel as fast as its weight and gravity force it to fall. It may land on a house roof, bounce off, and stop on the lawn.

These are all examples of terminal ballistics. They are varied and almost without end or number. Because of this, we must arbitrarily take certain specific situations and work with them. We do have many types of bullet design and construction in an attempt to provide sufficient variety to cover just about any potential situation. Unfortunately, it is usually difficult, if not impossible, to perceive a situation, study it, unload the handgun and reload with ammunition specifically designed for the situation at hand. In most instances, we must make do with what we have in the handgun when the situation arises, thus forcing officers to carry a compromise loading.

Such a compromise loading could work. It could also fail miserably. It could make a hero out of the officer or it could, just as easily, kill an innocent bystander and get the officer wounded. The ideal loading, of course, would force the bullet of large caliber to strike, enter, expand, and then, due to rapid and efficient expansion, come to a halt deep inside the felon driving him to the ground and rendering him incapable of further assault. If the distance is not too far and the officer's aim was good and the bullet placement was ideal and the felon was not under the influence of drugs, the combination of handgun, aim, and bullet design and construction should do the job.

Assume we have three cartridges, identical except for bullet construction. Each cartridge will drive its bullet at the same velocity as the other two. Each bullet has the same weight. If the first bullet strikes the target, due to its construction and design, it will instantly expand and go to pieces inflicting a surface wound. The second bullet, moving at the same velocity, fired from the same distance as the first bullet, strikes the target, drives right on through the target and continues on. The third bullet, under identical conditions, strikes the target, enters, expands, and stops within the target.

Obviously, the only bullet that imparts its full potential of impact shock into the target is the third bullet. The other two, while carrying the same potential impact shock, or foot pounds of energy, have failed to deliver their full potential into the target. This holds true of just this specific situation. Change any one factor in the target and the entire situation changes. Bullets may be designed and constructed to do a specific job within the limits of the ammunition capability. This is no great problem. The problem is to produce ammunition that will handle a wide variety of situations. This cannot be done; any ammunition will do one specific job better than others.

An extreme example would be to set up an orange on a post ten feet away. Load with 125 grain, .357 Magnum, high-velocity, hollow-point, jacketed ammunition. Taking careful aim, place the bullet into the orange. The orange will disappear in a hazy mist. When recovered, the bullet will be well expanded. Take the same ammunition and fire into a barricade of plywood, two by fours, and other materials, and the little 125 grain bullet will go to pieces with shallow penetration. This is an extreme example, but it points up the inability of ammunition to encompass the tremendous variety of situations that can arise.

One of the purposes of terminal ballistics is to evaluate bullet performance so that some indication of a bullet's potential may be understood and evaluated. If sufficient firing is conducted under controlled conditions, trends in bullet performance will appear and an understanding of a specific bullet's potential will become possible. If an organization has the financial backing, it can get into exotics such as the recent tests of ammunition conducted by the Bureau of Standards for LEAA, (see Chapter 16) but even such tests contain pitfalls for the unwary.

One of the most common methods of evaluating bullet performance is to fire it into a mass of papers or into a can of water catching the bullet in moist sawdust. The problem is that almost any bullet will expand when it strikes a gallon can full of water. It will also blow the can to pieces through hydraulic shock, not because of any great or unusual ability to deliver tremendous impact shock.

Paraffin and grease work well but are messy, troublesome, and time-consuming. One of the better constants is Duxseal held at temperatures around seventy degrees. It not only permits an examination of the wound channel but allows us to make plaster cases of this wound channel. It is not a substitute for living flesh, but it does retain the bullet; if the results are evaluated correctly, much can be learned regarding bullet performance.

Chapter 13

BULLET PERFORMANCE EVALUATION

THERE are many methods of attempting to evaluate bullet performance — wet papers, wet magazines, sand, clay, ordnance gelatin, a mixture of grease and paraffin, and Duxseal. The problem with most of these substances is that it becomes difficult to retain a constant consistency. Anything may be done if money is no object and there is sufficient labor to handle the problem at hand.

It is impossible to duplicate living flesh, bone, muscle. Nothing has yet been developed that will equal or duplicate these natural factors. Similarly, nothing has been able to include the human mind, adrenalin, fear, anger, intoxication, and the other stimuli that drive the living muscles and flesh to lengths commonly considered impossible when at rest.

Far too many people believe that when a bullet strikes, the victim drops in his or her tracks. This is not necessarily the case. A man sitting in a chair, reading a newspaper, quietly and leisurely having a drink before dinner may be killed instantly with a .22 caliber short. The same man with his adrenalin running wide open and stimulated by terrible anger can absorb half a dozen .45 bullets before dropping.

None of this can be readily built into a test target. Therefore, when we discuss bullet performance at the target, we must realize that the results are merely indicative of what could happen under certain conditions. Probably one of the finest test mediums is ordnance gelatin filled with bones of various sizes and diameters. It is seldom that a bullet will strike the human body and not strike a bone. On the other hand, any medium that will permit a wadcutter .38 Special bullet to produce higher impact shock and a larger cavity than a .45 ACP 230 grain bullet has definite disadvantages. Gelatin is extremely sensitive to bullet nose design so that an empirical evaluation is not always possible. Because of all these qualifications, I have used Duxseal for many years.

Duxseal made by Johns Mansville, provides a relative constant and it is available throughout the country, at plumbing supply houses. In addition, it is not necessary to heat, prepare, or spend time setting up for tests when using Duxseal. The bullet cavity or wound channel is available and plaster of paris casts may be made without any problems or the expenditure of large amounts of money.

Duxseal does have disadvantages. First, temperature is important because the consistency of Duxseal will vary with the temperature, thus giving differing results. As an example, if tests are run with the Duxseal at 68 degrees and then several days later confirming tests are run with the Duxseal at 79 degrees, results will not be consistent. Temperatures should be kept around 70 degrees. Also, Duxseal does tend to force a bullet to expand far more rapidly than the same bullet will expand in living flesh. However, I believe that through the use of recovered bullets and plaster of paris casts of the wound channel, an empirically correct bullet evaluation may be carried out. I consider Duxseal to usually provide about 40 percent more bullet performance than living flesh, bone, and muscle. If this is kept in mind, it is not too difficult to conduct meaningful bullet evaluation tests.

Far too often, tests are run at a distance of about ten feet. This is fine but it is very important to also determine what the same bullet will do when fired out of the same handgun at twenty-five yards and again at fifty yards. It would be foolish to adopt a bullet that performed perfectly at ten feet but failed at twenty-five yards. Most targets are not at ten feet. Twenty-five yards is a good average distance for handgun shooting.

A bullet does what is is designed to do providing the target make-up meets the specifications of the test material. This means that if you change one single factor, you alter the results. This is why most bullet tests bear little resemblance to what happens in the field. Another way of saying all this is to state that actually we do not know what will happen. On the other hand, we can try to provide both a projectile and a velocity at the target that should, under normal conditions, do the job at hand.

In the handgun game, we are not working with the wide

range of performance factors that we have in rifle ammunition. Generally speaking, we can take an average hunting rifle and its factory ammunition, and put down any reasonable game from fifty yards to four hundred yards, because we are working with high velocities plus extremely versatile bullet construction that will produce performance within a wide range of velocities. This is not the case with the handgun.

When using a machine rest, the Duxseal may be moved out to ever-increasing distances and bullets fired into it to determine bullet performance at different ranges. This is a slow process but one I consider essential to bullet evaluation. As I have already mentioned, both the wound channel and the bullet are instantly available (via plaster of paris casts), thus providing a much more critical determination of what may be expected of a specific bullet when fired from a specific handgun at a specific velocity. Obviously, changes in temperature must be kept in mind when discussing bullet performance.

Joyce Hornady wrote me recently: "We have done a lot of work in developing handgun bullets for expansion. For our expansion tests, we use a mixture of 80 percent paraffin wax and 20 percent cup grease melted together and cast in square bread pans about two-inches deep. These blocks are then stood on edge in a long box so a number of them makes a solid column of test media as long as necessary. The Duxseal we used to use was an ideal test medium except that it was just too tough. Everything expands beautifully in it whether it would expand in animal tissue or not. The material we now use is much closer to the resistance of animal tissue."

Remington told me some time ago that they use half-gallon milk cartons filled with water. Behind the cartons they have a large amount of cotton waste. The bullets are fired into the water-filled cartons to force expansion and then these same bullets are caught in the cotton waste.

As is obvious, no one has yet come up with the perfect material. I continue to use Duxseal because of ease of handling. Nothing has to be prepared; perhaps I am just lazy. I have found, over the years, that bullet expansion and penetration in Duxseal can be related to what a bullet should do in the field. By this I mean that if the tests indicate minimum expansion and maximum penetration that I can expect the same basic results in the field on game. Conversely, if bullets expand greatly and have only a very short wound channel, the same basic result should happen in the field. If you know that a bullet has certain tendencies when fired into Duxseal, you can learn what to expect from this bullet in the field.

Years ago, when the problem was how to get bullets to expand rapidly, we used oranges and grapefruit. A lightweight, high-velocity, expanding bullet would blow an orange or grapefruit into a cloud of haze. A round-nose, hard lead, slow-moving bullet would merely make a hole — in one side and out the other. We could then put a high-velocity expanding bullet into the fruit and blow it into millions of pieces.

All of these tests can be used to evaluate bullet performance. You will discover, over a period of time, that a relationship does exist between results in bullet testing and results in the field.

It is important that testing is not all done at close distances. How can you obtain any idea of what your bullet will do at eighty-five yards if all of your testing is done at ten feet? Try a few shots close up, then move your testing material out further and further. Bullet velocity and performance at the target can vary substantially from bullet velocity and performance at ten feet.

The type of big game you will be hunting can also affect the bullet performance you want. For example, a bullet that will drop a white tail may not even penetrate the wool and hide of a mountain sheep or goat. Keep all these thoughts in your mind as you test your ammunition. This testing is not something you can do in a few moments. It requires room, varying distances, and a facility where you can take your time without interruptions. Obviously, safety is an important factor.

In my opinion, anyone who plans to do any amount of handgun hunting should arrange for some place to fire tests and conduct evaluations of his ammunition. We have so much handgun ammunition on the market in so many different loadings that the handgunner cannot merely ask for a box of high velocity loadings. He must know the precise make and type of ammunition he will need for the job at hand. Similarly, the handloader must evaluate several different bullet weights, types, and makes.

The first thing I do is fire a bullet into Duxseal at about ten feet using the handgun with which I intend to hunt. If bullet expansion and penetration appear satisfactory, then I set this loading to one side and try another. As an example, the first loading may be Winchester 158 grain, flat-nose, jacketed, +P high-velocity .357 Magnum bullet. The next loading may be a similar Remington or Federal or Speer loading. If all produce acceptable performance, then I will use them in the final tests. If one loading appears to expand too rapidly, or fail to penetrate far enough, I may discard it at that point. Bear in mind that for game, we need a bullet that will drive down deeply into the animal to reach a vital organ, in addition to expanding somewhat.

On the other hand, a heavy, Keith-type hard lead bullet may penetrate ideally, but lack expansion; this may turn out to be precisely what we want for the game in mind. When we hunt deer, elk, moose, sheep, and goats, we must have a bullet that penetrates. Expansion can become secondary in this field.

Once I have selected a few loadings for actual testing, I set up my Ransom Machine Rest on my fifty-yard bench rest and set the handgun into the Ransom grip adapters. After firing several rounds to seat the handgun into the rest, I adjust point of bullet impact so that I can place a block of Duxseal in line with the point of bullet impact. Before actually firing into the Duxseal, I fire half a dozen of the loads to be tested to double check the precise point of bullet impact with the potential hunting load. Then I move the Duxseal into place and fire a single shot into it.

I remove the bullet and examine it carefully, noting the bullet channel. If I am basically satisfied with the results, I work the Duxseal back into a solid mass and fire another shot into it. I then turn the Duxseal so that the bullet entrance hole is up and pour it full of plaster of paris. After it hardens, I peel away the Duxseal and remove the cast with the bullet still affixed to the end of the case. This gives me a permanent record of what that load and bullet will do from that specific handgun at fifty yards. I then repeat this with all of the other potentially good loadings. The use of the Ransom rest eliminates missing the Duxseal or striking if off center. It is important, when using a Duxseal block, to place the bullets as close to the center as possible.

I do not bother to chronograph any of these loads because velocity has no affect upon the results, i.e. if bullet performance is good, velocity makes no difference. If we are working with close to maximum velocities, which is customary with hunting loads, velocity as such is not important. It is the bullet performance that is important. Obviously, there is no reason why velocities cannot be taken. If the shooter believes that velocities should be taken at the target, it is simple to set up an Oehler Sky Screen chronograph at the target. Once the point of bullet impact has been determined, move the chronograph holder into place, turn on and check the machine, and go back to the bench rest and fire. I just do not care that much about velocity at the target as long as the bullet is doing the required job.

BULLET EXPANSION

The following tests were conducted to illustrate a single method of evaluating bullet performance. Duxseal was used for capturing all bullets and two distances were used for firing, ten feet and twenty-five yards.

In an attempt to show the influence of barrel length upon bullet expansion, three barrel lengths were used: two-, four-, and six-inch. The handgun used was the Dan Wesson .357 Magnum.

Remington Ammunition	*Barrel Length*	*Distance 10 feet*	*Distance 25 yards*
125M1	2″	6.98	
	4″	5.62	6.32
	6″	5.98	
158M2	2″	6.61	
	4″	6.78	5.93
	6″	7.06	
158M3	2″	5.34	
	4″	5.89	5.57
	6″	6.44	
158M4	This bullet does not expand and is not designed for producing impact shock.		

158M5	2″	5.71	
	4″	6.13	5.67
	6″	6.23	
158 A	2″	5.59	
	4″	5.96	5.68
	6″	5.90	
125 B	2″	6.81	
	4″	6.52	6.58
	6″	5.79	
158 C	2″	5.70	
	4″	5.43	5.63
	6″	5.96	
110 D	2″	6.06	
	4″	6.76	6.12
	6″	7.57	

It is important to understand that the above figures represent basic averages in bullet expansion. One bullet may turn out quite round while another may have several partially torn off fragments. These fragments must not be measured; when such a bullet is recovered, another should be fired to obtain a reasonably normal amount of expansion. We are not attempting to impress anyone. We are only trying to discover certain basic characteristics of bullet performance to aid in evaluating such a bullet and load.

Figure 13-1. Remington M-1 bullets fired into Duxseal from (left) two-inch, (center) four-inch, and (right) six-inch Dan Wesson revolver at a distance of ten feet; (bottom) from a four-inch Dan Wesson revolver at a distance of twenty-five yards.

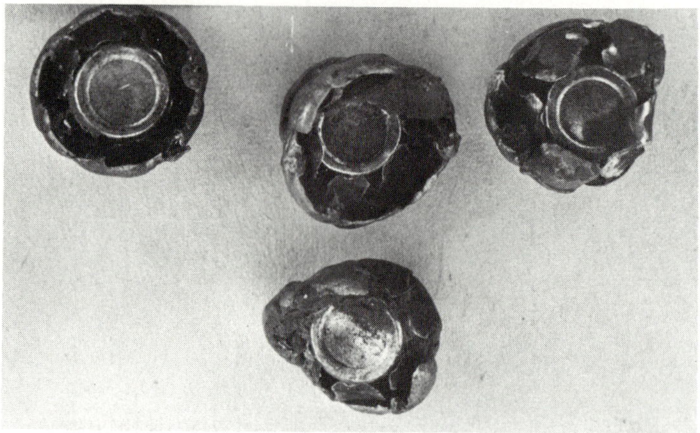

Figure 13-2. Remington M-2 bullets fired into Duxseal from (left) two-inch, (center) four-inch, and (right) six-inch Dan Wesson revolver at a distance of ten feet; (bottom) from a four-inch Dan Wesson revolver at a distance of twenty-five yards.

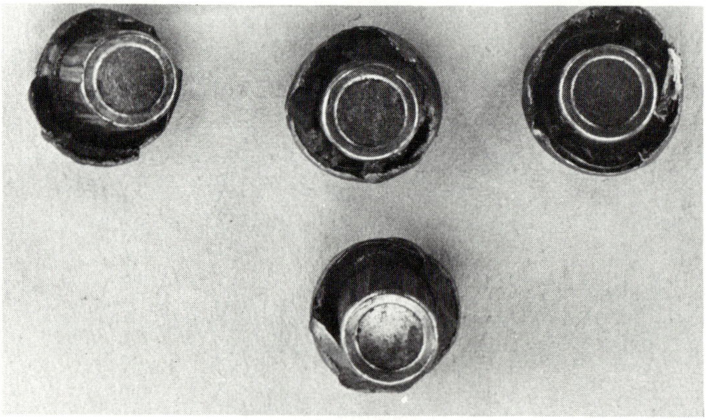

Figure 13-3. Remington M-3 bullets fired into Duxseal from (left) two-inch, (center) four-inch, and (right) six-inch Dan Wesson revolver at a distance of ten feet; (bottom) from a four-inch Dan Wesson revolver at a distance of twenty-five yards.

The figures illustrate the results. When studied along with the above data, a trend can often be seen.

It is important to also realize that no single test can correctly

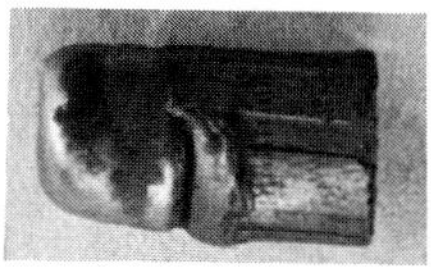

Figure 13-4. Remington M-4 cartridge fires a metal-point bullet that is not designed to expand; shown as recovered.

Figure 13-5. Remington M-5 bullets fired into Duxseal from (left) two-inch, (center) four-inch, and (right) six-inch Dan Wesson revolver at a distance of ten feet; (bottom) from a four-inch Dan Wesson revolver at a distance of twenty-five yards.

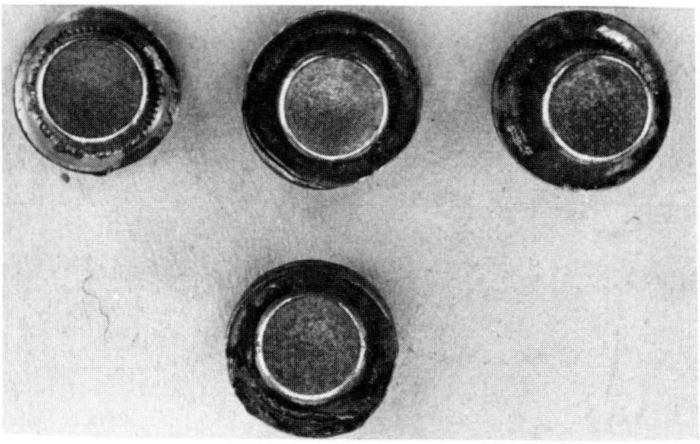

Figure 13-6. Federal A bullets fired into Duxseal from (left) two-inch, (center) four-inch, and (right) six-inch Dan Wesson revolver at a distance of ten feet: (bottom) from a four-inch Dan Wesson revolver at a distance of twenty-five yards.

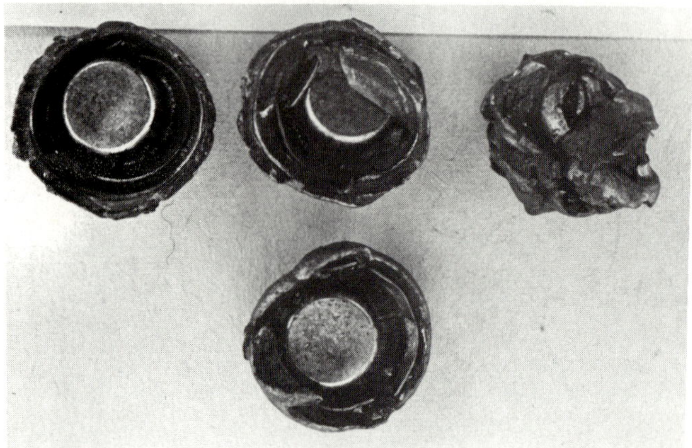

Figure 13-7. Federal B bullets fired into Duxseal from (left) two-inch, (center) four-inch, and (right) six-inch Dan Wesson revolver at a distance of ten feet; (bottom) from a four-inch Dan Wesson revolver at a distance of twenty-five yards.

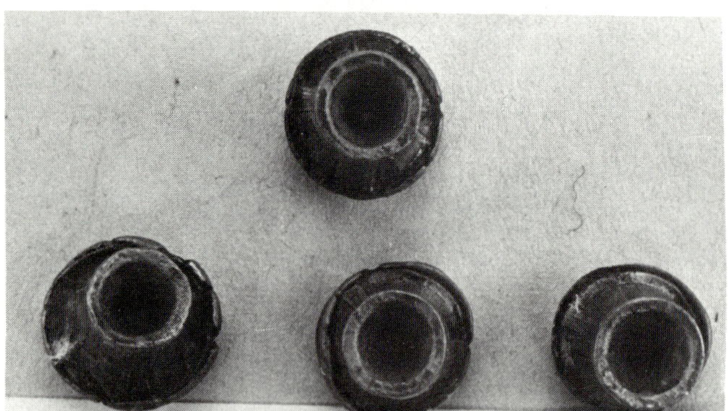

Figure 13-8. Federal C bullets fired into Duxseal from (left) two-inch, (center) four-inch, and (right) six-inch Dan Wesson revolver at a distance of ten feet; (bottom) from a four-inch Dan Wesson revolver at a distance of twenty-five yards.

evaluate bullet performance. Just because a bullet may produce excellent and desirable results in Duxseal may not indicate that it would be the ideal bullet. In another test it may fail to expand or it may go to pieces. When running tests on bullet

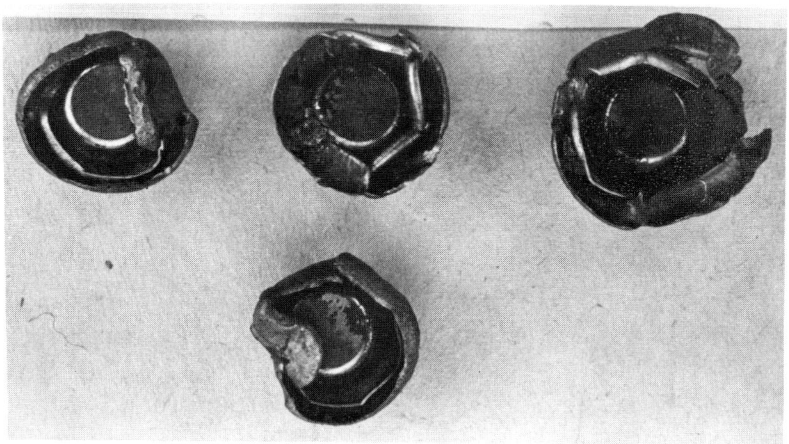

Figure 13-9. Federal D bullets fired into Duxseal from (left) two-inch, (center) four-inch, and (right) six-inch Dan Wesson revolver at a distance of ten feet; (bottom) from a four-inch Dan Wesson revolver at a distance of twenty-five yards.

performance, do not permit your desires to obtain the desired results to influence common sense. Weigh a series of perhaps eight or nine tests and then attempt to select the load and bullet that appears to do the job best.

TESTING BULLETS ON LIVING FLESH

Entirely aside from humanitarian considerations, the testing of bullets on living flesh is impractical. When I lived back east, one of my sources of relaxation was the hunting of woodchuck. Farmers went out of their way to cooperate with precision shooters because of the devastation and holes and lost feed attributable to the woodchuck. I have used just about every high-velocity .22 caliber cartridge produced. Over the years I came to the conclusion that it was not so much the cartridge or the velocity or the bullet weight or the accuracy but where the bullet hit that put down the woodchucks. Bear in mind that these varmints weighed around ten to twenty pounds. They were large and solid. Hit in the head, they went down. Hit elsewhere, sometimes they would go down; many times they would get deep into their holes.

It becomes obvious that it is impossible to take a series of identical living animals (identical in weight, mental condition, and physical condition) and place a series of identical bullets into these animals so that each bullet strikes its animal in precisely the same spot. It just cannot be done. It is possible to state that under normal conditions a high-velocity, 158 grain flat-nose, .357 Magnum bullet at a distance of thirty-eight yards will down a woodchuck more efficiently than a full-metal-jacket 124 grain 9 mm. loading. We can accept similar generalities but to state that one bullet is far more effective than another is often impossible.

Far too often the mental condition of the animal — whether man or beast — can control the effectiveness of any loading, with a highly stimulated animal failing to react as expected. Conversely, a dozing animal usually will never move after being struck by a far less efficient loading.

For these reasons, I cannot accept the many tables and calculations that have been turned out in the last twenty or thirty years. Tests made with gelatin, newspapers, mud, sand, butcher's meat, live goats, and all the other materials available do not necessarily reflect the capabilities of a bullet. These tests may indicate a trend or a basic result but any test that can place a .38 Special wadcutter, low-velocity lead bullet ahead of a 230 grain .45 ACP bullet obviously leaves much to be desired. All these tests lack the variables found in living animals: mental stimulus or the lack thereof, physical condition, distance from the handgun muzzle, residual velocity, and other factors that can control the effectiveness of the handgun projectile.

I could cite a long series of examples of bullet performance over the years but they could be entirely different than examples given by someone else. I agree that basically we know what is good, what is not so good, and what is ineffective. Far too many handgunners and officers completely fail to relate the ammunition they use to the situations they encounter. This must be corrected to reduce loss of life in law enforcement and the needless wounding of game in the field.

Chapter 14

FOOT POUNDS OF ENERGY

FOOT pounds of energy (FPE) is possibly one of the most used and most misunderstood theories in the history of firearms. Unfortunately, these FPE figures cannot be taken literally. They must be evaluated so that they bear some relation to factual bullet performance at the target. As an example, let us assume that a bullet has 450 FPE. This simply means that a specific bullet of a specific caliber moving at a specific velocity will, when it strikes, carry within it a potential of 450 FPE.

This is like saying that if a person wishes to go from his house out in the country into town he can take a vehicle. The vehicle can be a tractor, a truck, a coupe, a sedan, a two-door or a four-door or even a three-door vehicle. The potential is there. The person can take a vehicle, but which vehicle? Obviously, the one that will best do the job. The same applies to interpreting bullet performance at the target. Merely quoting FPE means nothing; however, generally speaking the larger the bullet diameter, the more effective the bullet *should* be against the target. Since the invention of the bullet, it has been generally accepted that, the larger the bullet diameter, the heavier the bullet will be and the more effective it will be against the target.

Bear in mind that 450 FPE can be reached by many bullets of varying weights, velocities, and calibers. Lightweight bullets may be rammed out at high velocities to reach 450 FPE. There is nothing magical about 450 FPE or any other figure. FPE is only a paper figure indicating that a projectile reaching this figure will have a potential of 450 FPE, regardless of whether or not this projectile delivers 450 FPE into the target.

The procedure for determining the foot pounds of energy of a projectile follows: Assume that we have a velocity of 1000 fps. Assume a bullet weight of 150 grains. Tables show that the key factor per grain of bullet weight is 2.22, so we multiply this key factor by the total bullet weight of 150 and arrive at a total potential of 333 FPE.

This time, assume we have a .45 ACP firing a 230 grain bullet at 800 fps. The key factor at this velocity is 1.42. Multiply by 230 (bullet weight) and we have an FPE figure of 326.0. We can also use a lightweight bullet; a 110 grain bullet at 1500 fps gives 550 FPE. The novice would immediately jump to the conclusion that the 110 grain bullet load is "fantastic." These FPE figures can be so deceptive that men are sent to their deaths, literally, because ammunition performance is predicated upon FPE and not upon empirical factors.

Bullet performance depends primarily upon three factors: bullet velocity, bullet construction, and the make-up of the target, (the composition of the target at the precise point of bullet impact). If a bullet strikes flesh and nothing else, it will perform entirely differently than if it strikes flesh, bone, muscle.

Bullet velocity is, obviously, a necessity. A stationary bullet cannot inflict damage upon anything. It must move in order to reach the target. Generally speaking, the higher the velocity upon impact, the more damage the bullet will do. We have all heard of a straw blown through a telephone pole during a hurricane. Velocity and velocity alone accomplished this. Basically we must have reasonable velocity in order to obtain reasonable bullet performance at the target.

If all we needed were velocity and the ability of the bullet to penetrate, then we could use pointed, full-metal-jacketed, metal-penetrating bullets. Such a bullet would have very limited practical use and be extremely dangerous in urban and suburban areas. Furthermore, while carrying perhaps a high FPE figure, it would impart little of this potential into the target.

Another bullet of perhaps the same weight and moving at the same velocity could, at the other extreme, be so constructed that it would expand and go to pieces upon impact without being able to penetrate the target, thus inflicting a surface wound. Again, while carrying a high FPE figure, it would not deliver this potential into the target.

A third bullet of the same weight moving once again at the

same velocity could be so designed and constructed that it would strike, penetrate, and then expand more and more rapidly as it moved within the target, thus delivering all of its FPE into the target. Obviously, such a bullet would be far more practical and deliver far more impact shock into the target than either of the other two.

Bullet construction is possibly more important than any other aspect, although there must be a definite and deliberate relation between bullet construction and the target at which the bullet must be fired.

For example, if most targets will be people, then certain predetermined characteristics may be designed into a bullet. True, such a bullet might also be acceptable for use against glass, cars, thick doors, walls, and similar obstacles, but these characteristics would be secondary to the primary purpose of putting down felons. It is not difficult to design and construct a bullet to do a basic job; it is difficult to do so and still provide sufficient secondary characteristics so that the bullet may be used effectively against other targets. In addition, the final bullet design should reduce or even eliminate ricochets.

A classic example of target construction occurred many years ago when officers cornered a felon in a bathroom. He took shelter within an old, heavy steel shower stall. The officers, firing round-nose, 158 grain lead bullets at low velocities, were unable to do more than dent the shower stall. It was a stand off until the officers brought in tear gas.

There are times when even the most thoroughly tested ammunition will fail to do the job. Not too many years ago, an officer answered a call in an urban area. He went up the first flight of narrow stairs into the second floor hallway, and just as he started up the second flight of stairs, his quarry appeared at the top of the stairs. The two men eyed each other for a moment. The officer was well built, young, an athlete, and in good condition. The other man stood six foot four, weighed over three hundred pounds, held an iron bar in his right hand, and appeared to be under the influence of liquor. The officer drew his handgun, a .38 Special loaded with 158 grain, jacketed, soft-point, flat-nose cartridges, and ordered the man to put down the iron bar. The man flung himself down the stairs at a run, raising the iron bar to strike the officer. The officer emptied his handgun into the vast bulk of rapidly moving flesh and jumped back to one side. Silently the man slammed into the tenement wall, turned, and started beating the officer with the iron bar. Only the arrival of back-up officers saved the first officer.

As may be seen, six shots emptied into a man with perfect terminal ballistical results at close range do not necessarily mean the end of a situation.

This is an ideal example of how important target construction can become. As a result, FPE can be a guide to bullet performance but only if correctly evaluated. Such an evaluation must take into consideration what the basic targets will be. Then the correct bullet design and construction must be determined, and it will be necessary to find ammunition loaded with that basic bullet design in the needed calibers. Quite often an FPE of 300 may be far more practical than an FPE of 400.

As may be seen, evaluating bullet performance is a long and detailed task. It cannot be done by running a finger down a table and arbitrarily selecting a predetermined FPE.

Chapter 15

THE GLASSER SAFETY SLUG

THE concept of the multiple projectile is not new either militarily or medically. The theory is simple and practical. Assume we have a single projectile weighing 160 grains. This projectile will deliver 355 foot pounds of energy into a target under correct conditions if traveling at a velocity of 1,000 fps. when it strikes the target.

Now, supposing in place of a single projectile, we use two projectiles, each weighing 80 grains. Velocity is the same, as is everything else. What happens? We obtain substantially more total impact shock from two projectiles than we do from a single projectile of the same total weight. These two projectiles strike one after another. They do not impact at precisely the same time. There is a slight time lapse. Despite the fact that the total weight remains 160 grains, the actual impact shock delivered over a longer period of time than that of a single 160 grain projectile substantially increases the shock delivered into the target. The longer the duration of the impact shock, the more devastating the results. Bear in mind that we are dealing with milliseconds. Put another way, impact shock delivered over a period of one-tenth of a second will not be as potent as impact shock delivered over a period of one-seventh of a second.

The development of the Glasser Safety Slug* is based upon this concept of extended impact shock. Many problems arose during the initial stages of development of a practical projectile but today the Glasser Slug is an accepted and established projectile that is loaded into many handgun calibers, among them the 9 mm Luger, .38 Special, .45 ACP, .44 Special, .44 Magnum, and .357 Magnum.

It is impractical for self-defense or hunting purposes or most law enforcement work to carry a fully automatic weapon. The

*Invented by Mr. Glasser, Switzerland. Manufactured by Jack Cannon, P.O. Box 1975, McAllen, Texas 78501.

advantage of a fully automatic piece is that a spray of projectiles may be dumped into the target, thus putting it down without any doubt; this is one of the basic reasons for the adoption of the M-16 .223 caliber rifle and cartridge. A single .223 bullet, weight 55 grains, will seldom put down and stop an attacker. A burst of fifteen of these bullets placed into the attacker will dispose of him. In other words, a multitude of projectiles has replaced precision firing with a single projectile. This concept is not suited for most civilian and law enforcement situations. The Glasser Slug approaches this problem by providing the handgunner with a single, multi-projectile projectile. If the shooter can place the single projectile into the target, his worries are over because the multi-projectiles contained within the single bullet produce a long impact shock plus devasting results.

This is accomplished by encasing #12 shot within a conventional bullet jacket and then fusing the two into a single unit by means of a liquid plastic compound that, upon drying, becomes frangible. Once this plastic has set within the bullet jacket, a nose-forming and core-protection coat of a different plastic is formed over the bullet nose to produce the final projectile. This composite projectile is now ready to be seated into the cartridge case.

These bullets are lighter in weight than conventional bullets of the same size, therefore, velocities are quite high for handgun ammunition. When these bullets strike the target, it takes a moment for the nose cap of plastic to move back and shatter both the inner plastic core of plastic filled with #12 shot and the conventional bullet jacket. The moment the Glasser Slug starts to break up, the impact shock is tremendous.

Because of the slight delay in breaking up, the Glasser Slug has far more penetration potential than expected. On the other hand, the moment the bullet commences to break up, penetration ceases; thus, the full force of the impact shock is always delivered within the target. As a result, this projectile is almost ricochet proof. Once it strikes a hard object, such as concrete or a brick wall or similar obstruction, it goes to pieces and cannot ricochet.

As may be seen, the length of time between the moment the

bullet strikes the target and the time the large number of #12 shot stops within the target is extremely long for a high velocity projectile, thus producing a tremendous number of minor impact shocks over a long period of time.

Empirically, performance is ideal and the tremendous safety factor of low penetration and lack of ricochets make such a loading ideal for many purposes. It will be found in increasing numbers because, in the near future, it will be sold to civilians. In the past, sales were strictly limited to law enforcement personnel due to the inability to produce the bullets on a mass production basis; production difficulties have now been solved.

As any student of ballistics will instantly notice, two things come to mind. First, the use of a Glasser Safety Slug upon a target will be obvious. Next, in most instances, it is impossible to recover sufficient jacket material to enable a ballistician to match the projectile to a specific barrel.

On the other hand, the efficiency and accuracy at handgun ranges make this loading extremely desirable for law enforcement work. When tied in with its inherent safety, it is difficult to imagine a more ideally designed law enforcement cartridge.

Chapter 16

POLICE HANDGUN AMMUNITION

The complete text of the bulletin "An Evaluation of Police Handgun Ammunition may be obtained from the Law Enforcement Standards Laboratory, National Bureau of Standards, U.S. Department of Justice, Washington, D.C.

The final report was issued in 1978 and should be read and studied by all personnel involved in the selection of handgun ammunition.

I would like to inject a word of caution. Far too many people with a moderate knowledge of ballistics and ammunition will tend to accept these tables and this data as verbatim. Unfortunately, this should not be done. Use all this data as a guide, as a reference point from which additional and specific tests may be conducted. As an example, the following tables indicate that a .38 Special, wadcutter, lead bullet loading with a velocity of around 750 fps has more stopping power than a .45 ACP round-nose, full-metal-jacketed, 230 grain bullet load. Empirically, this is wrong, despite the fact that the tests showed that such is the case. This came about through the use of gelatin compound that reacted violently to the impact of the sharp-shoulder, flat-nose, .38 Special wadcutter bullet. When the .45 full-metal-jacket bullet struck the gelatin, it wormed its way through the gelatin with only a slight disturbance.

For these and similar reasons, any set of tables must be examined carefully. Never accept any such tables or tests as a final, indisputable fact. There is no such thing in ballistics work. It is irresponsible for anyone to point to a specific loading and state that "this is the one and only loading that may be used because it says right here in the table that this loading is superior to another." It is obvious that common sense plus discussions with men who are specialists in these fields must play an important part in any ammunition evaluation.

* * *

TEST PROGRAM

AN extensive laboratory investigation of all significantly different handgun bullets in the caliber range 9 mm to .45, which are currently available to law enforcement agencies in the United States, has been essentially completed. These experiments include the following:

a. A determination of each bullet's behavior on striking and penetrating ordnance gelatin, as a function of its impact velocity.

b. Measurement of the formation and subsequent development of the temporary cavity produced in the gelatin by each projectile, using high-speed motion pictures.

c. Measurement of the dynamic behavior of each bullet as it penetrated the gelatin, i.e., its stability and deformation, using flash x-ray photography.

d. Measurement of the impact velocity of factory-loaded ammunition corresponding to each bullet under study, when fired from various handguns currently used by law enforcement agencies.

e. Measurements designed to determine the ricochet and penetration potentials of each bullet, as a function of angle of incidence and velocity, when striking various common materials.

There experiments were conducted in an attempt to determine what produces human incapacitation by handgun bullets (relative stopping power), which, if any, of the existing theories on this subject are correct and, in particular, whether relative incapacitation can be predicted solely in terms of the properties of the bullet, i.e., its mass, velocity, shape, construction, and caliber. An understanding and assessment of the relative hazards due to the ability of these bullets to ricochet from and to penetrate materials was also sought.

In addition to the experimental effort, theoretical models of cavity formation in gelatin by both deforming and nondeforming, nontumbling handgun bullets have been developed. These models have successfully predicted the observations made experimentally.

RELATIVE INCAPACITATION (RELATIVE STOPPING POWER)

The model for relative incapacitation developed in this study is based on a complete assessment of the confrontation between a law enforcement officer and an armed criminal. The behavior of a bullet in the target is only a part of this model. The vulnerability of humans to incapacitation by handgun bullets, the path of the bullet within the target, the point of aim, and the ability of the officer to hit his target are other important aspects of the general problem and have been considered in detail.

The incapacitation criterion imposed on the model is a stringent one, as it must be if meaningful estimates are to be made. To identify those parts of the human anatomy which, when rendered temporarily or permanently non-functional, will result in the instantaneous incapacitation of the criminal, the following scenario was presented to a team of medical experts:

> An armed criminal has been placed in a high-stress situation in which he is convinced that only continued aggression on his part will prevent the loss of his life; he is armed with a hand-held weapon and he is being approached by an armed law enforcement officer; he attacks the officer. The officer must administer a swift, incapacitating injury to the assailant with his handgun, in order to save his own life.

The medical experts were asked to assign relative values to those parts of the body whose wounding would produce this instantaneous incapacitation. Their judgments were further constrained by a definition of instantaneous incapacitation:

> Instantaneous incapacitation is that which will render the assailant incapable of posing a continued threat to the safety of the officer by use of a hand-held weapon. Such injury may include clinical death, unconsciousness, biomedical dysfunction, etc., but pain may not be considered a contributing factor.

These data have been encoded for use in conjunction with an

elaborate three-dimensional computerized model of the human anatomy, known as the Computer Man. The Computer Man is the result of several years of effort on the part of personnel at the Ballistic Research Laboratories, Aberdeen Proving Ground, and has proven to be a valuable tool in their overall program in wound ballistics.

The Computer Man consists of volume elements of the body of a man in the form of rectangular sections 5 × 5 × 25 mm (0.2 × 0.2 × 1 in) in size. Within each volume element, all tissue types have been identified and encoded. The composite of the above injury criterion (instantaneous incapacitation), the Computer Man, and the probable distribution of the impact points and trajectories of the bullets from a given handgun in terms of the range to the target and the point of aim, all enter into the assessment of the vulnerability of a human target to incapacitation by a particular handgun bullet type.

The feature of a bullet's interaction with soft tissue that has been identified as contributing most to instantaneous incapacitation is the temporary cavity which the bullet generates as it penetrates. This one feature has the most likelihood of interacting with those relatively few centers of vital activity required for continued aggression. For a given impact point, the bullet producing the largest temporary cavity at the proper depth of penetration — defined by the location of vital organs — will have the greatest chance for success in producing instantaneous incapacitation. In other words, such a round will possess the greatest stopping power.

The radius of the maximum temporary cavity, as a function of depth of penetration, has been chosen as the measure of *relative bullet performance.* This performance can now be predicted for several types of bullets, without recourse to experiment, because of our present ability to calculate these cavities using solely the velocity, shape, construction, mass and caliber of the bullet.

However, the *relative incapacitation potential* of a given bullet type will also depend on the point of aim and the probabilities of the bullet impacting the target at various points surrounding the point of aim. This latter is in turn dependent on the range, the marksmanship of the shooter and the characteristics of the bullet/weapon combination.

The present study has generated a very large volume of data which relate the maximum temporary cavity generated by a handgun bullet to its velocity, shape, construction, mass and caliber. These data are still being analyzed. However, progress to date does permit certain tentative conclusions to be drawn; these may provide some immediate assistance to law enforcement agencies now involved in ammunition selection.

CONCLUSIONS

In the range of calibers studied, the most important property of a moving handgun bullet affecting its performance in the target medium is its velocity. There are several reasons:

1. The size of the Maximum Temporary Cavity (MTC) depends partly on the striking kinetic energy, $1/2mv_o^2$, i.e., the volume of the MTC depends on the total energy available.

2. There is a threshold velocity, below which a bullet will not deform; deformation of the bullet greatly affects the size and shape of the MTC.

It should be stressed, however, the one cannot use the striking kinetic energy as the sole criterion for ranking handgun bullets. It is the size and shape of the resulting MTC and how it overlaps vital organs that ultimately gives one bullet a higher relative incapacitation index (RII) than another. Some lighter bullets yield a higher RII than heavier ones having the same striking kinetic energy, shape, construction and caliber.

TABLE 16-I

Performance of Commercially Available Handgun Ammunition

BULLET ID NO.	CALIBER	WEIGHT (grains)	BULLET TYPE	MANUFACTURER	BARREL LENGTH (in)	VELOCITY NOMINAL* (fps)	VELOCITY MEASURED (fps)	VELOCITY MEASURED (mps)	RI INDEX
1	.44 MAG	200	JHP	SPEER	4.00	1675	1277	389	54.9
2	9MM	96	SAFETY SLUG	DEADEYE ASSOC	4.00	1365	1839	560	54.5
3	.41 MAG	210	JSP	REMINGTON	4.00	1500	1260	384	51.9
4	.357 MAG	96	SAFETY SLUG	DEADEYE ASSOC	4.00	1120	1725	525	50.0
5	.44 MAG	240	SWC	WINCH-WESTERN	4.00	1470	1330	405	50.0
6	.44 MAG	240	SWC	BROWNING	4.00	1470	1311	399	49.8
7	.44 MAG	240	SWC	REMINGTON	4.00	1470	1286	391	48.9
8	.44 MAG	240	JHP	BROWNING	4.00	1330	1257	383	47.9
9	.44 MAG	240	JHP	REMINGTON	4.00	1470	1229	374	46.7
10	.357 MAG	96	SAFETY SLUG	DEADEYE ASSOC	2.00	1120	1615	492	46.0
11	.44 MAG	240	JSP	SPEER	4.00	1650	1203	366	45.7
12	.357 MAG	125	JHP	SPEER	4.00	1900	1301	396	44.4
13	.357 MAG	140	JHP	SPEER	4.09	1780	1221	372	44.4
14	.357 MAG	125	JHP	REMINGTON	4.00	1675	1366	416	42.5
15	.38 SPEC	96	SAFETY SLUG	DEADEYE ASSOC	4.00	1800	1585	483	41.8
16	.44 MAG	180	JSP	SUPER VEL	4.00	1995	1495	455	41.6
17	9MM	115	JHP	REMINGTON	4.00	1160	1192	363	38.0
18	.38 SPEC	96	SAFETY SLUG	DEADEYE ASSOC	2.00	1800	1496	455	37.5
19	.357 MAG	125	JHP	REMINGTON	2.00	1675	1173	357	37.1
20	.357 MAG	140	JHP	SPEER	2.00	1780	1125	342	34.4
21	.357 MAG	110	JHP	SPEER	4.00	1700	1246	379	33.4
22	.357 MAG	125	JHP	SPEER	2.00	1900	1161	353	30.6

BULLET ID NO.	CALIBER	WEIGHT (grains)	BULLET TYPE	MANUFACTURER	BARREL LENGTH (in)	VELOCITY NOMINAL* (fps)	VELOCITY MEASURED (fps)	VELOCITY MEASURED (mps)	RI INDEX
23	.357 MAG	158	JSP	SPEER	4.00	1625	1156	352	28.0
24	.38 SPEC	95	JHP(+P)	REMINGTON	4.00	985	1187	361	28.0
25	9MM	100	JHP	SPEER	4.00	1315	1188	362	27.9
26	.38 SPEC	125	JHP	REMINGTON	4.00	1160	1108	337	25.5
27	.38 SPEC	110	JHP	SUPER VEL	4.00	1370	1159	353	25.1
28	.38 SPEC	110	JHP	SUPER VEL	2.00	1370	1148	349	24.8
29	.357 MAG	110	JHP	SMITH+WESSON	4.00	1800	1226	373	24.0
30	.357 MAG	110	JHP	SPEER	2.00	1700	1178	359	23.3
31	.38 SPEC	125	JSP(+P)	SPEER	4.00	1425	1047	319	22.5
32	.357 MAG	125	JHP	SMITH +WESSON	4.00	1775	1227	373	22.1
33	.357 MAG	158	JSP(HI-VEL)	FEDERAL	4.00	1550	1255	382	21.1
34	.45 AUTO	185	JHP	REMINGTON	5.00	950	895	272	21.1
35	.357 MAG	110	JHP	WESTERN SUP-X	4.00	1500	1309	398	21.0
36	.357 MAG	110	JHP	WESTERN SUP-X	2.75	1500	1258	383	20.2
37	.38 SPEC	125	JHP(+P)	SPEER	4.00	1425	1006	306	19.9
38	.38 SPEC	90	MP	KTW	4.00	1030	922	281	19.6
39	.38 SPEC	110	JSP	SUPER VEL	4.00	1370	1202	366	19.4
40	.38 SPEC	110	JHP(LOT-Q4070)	WINCH-WESTERN	4.00	####	1106	337	19.3
41	.357 MAG	158	JSP(HI-VEL)	FEDERAL	2.00	1550	1195	364	18.7
42	.38 SPEC	140	JHP(+P)	SPEER	4.00	1200	978	298	18.6
43	.38 SPEC	140	JHP(+P)	SPEER	2.00	1200	897	273	18.5
44	.38 SPEC	158	LHP	WINCH-WESTERN	4.00	855	915	278	18.4
45	.357 MAG	125	JHP	SMITH+WESSON	2.00	1775	1188	362	17.7
46	.357 MAG	158	JSP	SPEER	2.00	1625	1030	313	17.5
47	.357 MAG	158	JSP	SMITH+WESSON	4.00	1500	1168	356	17.2
48	.357 MAG	158	JSP	SMITH+WESSON	2.00	1500	1091	332	17.0
49	9MM	115	JHP	SMITH+WESSON	4.00	1145	1193	363	16.6

BULLET ID NO.	CALIBER	WEIGHT (grains)	BULLET TYPE	MANUFACTURER	BARREL LENGTH (in)	VELOCITY NOMINAL* (fps)	MEASURED (fps)	MEASURED (mps)	RI INDEX
50	.357 MAG	158	LRN(LUBALOY)	WESTERN SUP-X	4.00	1410	1230	374	16.6
51	.38 SPEC	125	JSP	3-D	4.00	1085	1091	332	16.5
52	.38 SPEC	90	MP	KTW	2.00	1030	734	223	15.6
53	.38 SPEC	125	JHP(+P)	SPEER	2.00	1425	931	283	15.5
54	9MM	100	FJ(FMC)	SMITH+WESSON	4.00	1250	1341	408	15.2
55	.45 AUTO	185	WC(TARGETMASTER)	REMINGTON	5.00	775	821	250	14.7
56	.38 SPEC	125	JSP	SMITH+WESSON	4.00	1350	1064	324	14.5
57	.357 MAG	158	JHP	SMITH+WESSON	4.00	1050	1116	340	14.4
58	.357 MAG	158	LRN(LUBALOY)	WESTERN SUP-X	2.00	1410	1169	356	14.4
59	.38 SPEC	158	SWC	WINCHESTER	4.00	855	924	281	14.3
60	.38 SPEC	95	JHP(+P)	REMINGTON	2.00	985	1019	310	14.0
61	.38 SPEC	110	JHP(LOT-Q4070)	WINCH-WESTERN	2.00	####	956	291	14.0
62	.38 SPEC	110	JSP	SUPER VEL	2.00	1370	1076	327	14.0
63	.357 MAG	110	JHP	SMITH+WESSON	2.00	1800	1044	318	13.9
64	9MM	124	FJ(FMC)	REMINGTON	4.00	1120	1084	330	13.8
65	.41 MAG	210	SWC	REMINGTON	4.00	1050	944	287	13.7
66	.38 SPEC	125	JSP(+P)	SPEER	2.00	1425	983	299	13.2
67	.38 SPEC	158	JHP	SMITH+WESSON	4.00	1050	1047	319	13.0
68	.38 SPEC	90	JSP(HEMI)	SMITH+WESSON	4.00	1350	1158	352	12.4
69	.38 SPEC	110	JHP	SMITH+WESSON	4.00	1380	1014	309	12.4
70	.38 SPEC	148	WC	REMINGTON	4.00	770	741	225	12.4
71	.38 SPEC	148	WC	BROWNING	4.00	770	731	222	12.3
72	.38 SPEC	148	WC	FEDERAL	4.00	770	737	224	12.3
73	.38 SPEC	148	WC	SMITH+WESSON	4.00	800	726	221	12.3
74	.38 SPEC	148	WC	REMINGTON	2.00	770	700	213	12.2
75	.38 SPEC	148	WC	FEDERAL	2.00	770	674	205	12.1
76	.38 SPEC	148	WC	SMITH +WESSON	2.00	800	662	201	12.1

BULLET ID NO.	CALIBER	WEIGHT (grains)	BULLET TYPE	MANUFACTURER	BARREL LENGTH (in)	VELOCITY NOMINAL* (fps)	VELOCITY MEASURED (fps)	VELOCITY MEASURED (mps)	RI INDEX
77	.38 SPEC	148	WC	SPEER	4.00	825	679	206	12.1
78	.38 SPEC	148	WC(CLEAN CUTTING)	WESTERN	4.00	770	696	212	12.1
79	9MM	115	JSP(POWER POINT)	WESTERN SUP-X	4.00	1160	1272	387	12.0
80	.38 SPEC	148	WC	SPEER	2.00	825	652	198	12.0
81	.38 SPEC	148	WC	BROWNING	2.00	770	618	188	11.9
82	.38 SPEC	148	WC(CLEAN CUTTING)	WESTERN	2.00	770	618	188	11.9
83	.38 SPEC	90	JSP	SMITH+WESSON	4.00	1350	1118	340	11.8
84	.357 MAG	158	JHP	SMITH+WESSON	2.00	1050	982	299	11.1
85	.38 SPEC	158	LHP	WINCH-WESTERN	2.00	855	805	245	11.0
86	.38 SPEC	158	SWC	FEDERAL	4.00	855	823	250	10.9
87	.38 SPEC	158	SWC	SMITH+WESSON	4.00	850	1006	306	10.8
88	.38 SPEC	158	JHP	SMITH+WESSON	2.00	1050	950	289	10.6
89	.38 SPEC	110	JHP	SPEER	4.00	1245	857	261	10.5
90	9MM	115	FJ(FMC)	SMITH+WESSON	4.00	1145	1192	363	10.3
91	.357 MAG	158	SWC	REMINGTON	4.00	1410	1088	331	10.2
92	.38 SPEC	125	JSP	3-D	2.00	1085	957	291	10.1
93	9MM	125	JSP	SPEER	4.00	1120	1058	322	9.9
94	9MM	115	FJ(FMC)	WINCHESTER	4.00	1140	1126	343	9.7
95	.45 AUTO	185	WC	FEDERAL	5.00	775	751	228	9.7
96	.38 SPEC	125	JHP	SMITH+WESSON	2.00	1350	899	274	9.6
97	.357 MAG	158	SWC	REMINGTON	2.00	1410	958	291	9.3
98	9MM	115	FJ(FMC)	BROWNING	4.00	1140	1067	325	9.2
99	.38 SPEC	158	LRN(+P)	FEDERAL	4.00	1090	999	304	9.0
100	.38 SPEC	125	JHP	SMITH+WESSON	4.00	1350	1002	305	8.9
101	.38 SPEC	158	SWC	FEDERAL	2.00	855	796	242	8.5
102	.38 SPEC	158	SWC	SPEER	4.00	975	803	244	8.5
103	.38 SPEC	158	LRN(+P)	FEDERAL	2.00	1090	947	288	8.2

BULLET ID NO.	CALIBER	WEIGHT (grains)	BULLET TYPE	MANUFACTURER	BARREL LENGTH (in)	VELOCITY NOMINAL* (fps)	VELOCITY MEASURED (fps)	VELOCITY MEASURED (mps)	RI INDEX
104	.38 SPEC	158	SWC	WINCHESTER	2.00	855	799	237	8.2
105	.38 SPEC	158	LRN	WINCHESTER	4.00	855	919	280	8.0
106	.38 SPEC	110	JHP	SPEER	2.00	1245	789	240	7.7
107	.38 SPEC	90	JSP(HEMI)	SMITH+WESSON	2.00	1350	1053	320	7.2
108	.38 SPEC	125	JHP	REMINGTON	2.00	1160	911	277	7.0
109	.38 SPEC	110	JHP	SMITH+WESSON	2.00	1380	888	270	6.8
110	.45 AUTO	230	FJ	REMINGTON	5.00	855	839	255	6.7
111	.45 LC	225	LRN	WINCH-WESTERN	7.50	860	821	250	6.6
112	.38 SPEC	90	JSP	SMITH+WESSON	2.00	1350	975	297	6.5
113	.45 AUTO	230	FJ	WINCH-WESTERN	5.00	850	740	225	6.5
114	.44 SPEC	246	LRN	REMINGTON	2.00	755	640	195	6.3
115	.38 SPEC	125	JHP	SMITH+WESSON	4.00	1350	900	274	5.9
116	.38 SPEC	158	SWC	SPEER	2.00	975	640	195	5.7
117	.38 SPEC	125	JSP	SMITH+WESSON	2.00	1350	896	273	5.6
118	.38 SPEC	158	LRN	FEDERAL	4.00	855	795	242	5.0
119	.38 SPEC	158	LRN	WINCHESTER	2.00	855	780	237	4.6
120	.38 SPEC	158	LRN	REMINGTON	4.00	855	749	228	4.5
121	.38 SPEC	158	LRN	SPEER	4.00	975	749	228	4.5
122	.38 SPEC	200	LRN	REMINGTON	4.00	730	647	197	4.5
123	.38 SPEC	200	LRN	SPEER	4.00	850	710	216	4.5
124	.38 SPEC	158	LRN	REMINGTON	2.00	855	694	211	4.4
125	.38 SPEC	158	LRN	SPEER	2.00	975	635	193	4.4
126	.38 SPEC	158	LRN	SMITH+WESSON	4.00	910	708	215	4.4
127	.38 SPEC	158	LRN	FEDERAL	2.00	855	632	192	4.2
128	.38 SPEC	200	LRN(LUBALOY)	WESTERN SUP-X	4.00	730	626	190	4.2
129	.38 SPEC	200	LRN	SPEER	2.00	850	598	182	4.1
130	.38 SPEC	200	LRN(LUBALOY)	WESTERN SUP-X	2.00	730	592	180	4.1

BULLET ID NO.	CALIBER	WEIGHT (grains)	BULLET TYPE	MANUFACTURER	BARREL LENGTH (in)	VELOCITY NOMINAL* (fps)	VELOCITY MEASURED (fps)	VELOCITY MEASURED (mps)	RI INDEX
131	.38 SPEC	158	SWC	SMITH+WESSON	4.00	1060	875	266	4.0
132	.38 SPEC	158	SWC	SMITH+WESSON	2.00	850	870	265	4.0
133	.38 SPEC	200	LRN	REMINGTON	2.00	730	593	180	4.0
134	380 AUTO	95	FJ	WESTERN SUP-X	3.86	955	948	288	4.0
135	.38 SPEC	158	LRN	SMITH+WESSON	2.00	910	626	190	3.5
136	.38 SPEC	125	JHP	SMITH+WESSON	2.00	1350	716	218	3.0
137	.38 SPEC	158	JSP	SMITH+WESSON	4.00	1050	828	252	2.9
138	.38 SPEC	158	SWC	SMITH+WESSON	2.00	1060	678	206	2.5
139	.22 CAL	37	LHP	WINCH-WESTERN	2.00	1365	872	265	2.3
140	.38 SPEC	158	JSP	SMITH+WESSON	2.00	1050	730	222	2.0
141	.38 SPEC	64	SHORT STOP	MBA	4.00	####	738	224	0.9
142	.38 SPEC	64	SHORT STOP	MBA	2.00	####	671	204	0.4

* = Advertized Velocity

= Velocity not available

TABLE 16-II

Performance of Commercially Available Handgun Ammunition

BULLET ID NO.	CALIBER	WEIGHT (grains)	BULLET TYPE	MANUFACTURER	BARREL LENGTH (in)	VELOCITY NOMINAL* (fps)	VELOCITY MEASURED (fps)	VELOCITY MEASURED (mps)	RI INDEX
139	.22- CAL	37	LHP	WINCH-WESTERN	2.00	1365	872	265	2.3
2	9MM	96	SAFETY SLUG	DEADEYE ASSOC	4.00	1365	1839	560	54.5
54	9MM	100	FJ(FMC)	SMITH+WESSON	4.00	1250	1341	408	15.2
25	9MM	100	JHP	SPEER	4.00	1315	1188	362	27.9
98	9MM	115	FJ(FMC)	BROWNING	4.00	1140	1067	325	9.2
90	9MM	115	FJ(FMC)	SMITH+WESSON	4.00	1145	1192	363	10.3
94	9MM	115	FJ(FMC)	WINCHESTER	4.00	1140	1126	343	9.7
17	9MM	115	JHP	REMINGTON	4.00	1160	1192	363	38.0
49	9MM	115	JHP	SMITH+WESSON	4.00	1145	1193	363	16.6
79	9MM	115	JSP(POWER POINT)	WESTERN SUP-X	4.00	1160	1272	387	12.0
64	9MM	124	FJ(FMC)	REMINGTON	4.00	1120	1084	330	13.8
93	9MM	125	JSP	SPEER	4.00	1120	1058	322	9.9
10	.357 MAG	96	SAFETY SLUG	DEADEYE ASSOC	2.00	1120	1615	492	46.0
4	.357 MAG	96	SAFETY SLUG	DEADEYE ASSOC	4.00	1120	1725	525	50.0
29	.357 MAG	110	JHP	SMITH+WESSON	4.00	1800	1226	373	24.0
63	.357 MAG	110	JHP	SMITH+WESSON	2.00	1800	1044	318	13.9
21	.357 MAG	110	JHP	SPEER	4.00	1700	1246	379	33.4
30	.357 MAG	110	JHP	SPEER	2.00	1700	1178	359	23.3
35	.357 MAG	110	JHP	WESTERN SUP-X	4.00	1500	1309	398	21.0
36	.357 MAG	110	JHP	WESTERN SUP-X	2.75	1500	1258	383	20.2

BULLET ID NO.	CALIBER	WEIGHT (grains)	BULLET TYPE	MANUFACTURER	BARREL LENGTH (in)	VELOCITY NOMINAL* (fps)	VELOCITY MEASURED (fps)	VELOCITY MEASURED (mps)	RI INDEX
32	.357 MAG	125	JHP	SMITH+WESSON	4.00	1775	1227	373	22.1
45	.357 MAG	125	JHP	SMITH+WESSON	2.00	1775	1188	362	17.7
12	.357 MAG	125	JHP	SPEER	4.00	1900	1301	396	44.4
22	.357 MAG	125	JHP	SPEER	2.00	1900	1161	353	30.6
14	.357 MAG	125	JHP	REMINGTON	4.00	1675	1366	416	42.5
19	.357 MAG	125	JHP	REMINGTON	2.00	1675	1173	357	37.1
13	.357 MAG	140	JHP	SPEER	4.00	1780	1221	372	44.4
20	.357 MAG	140	JHP	SPEER	2.00	1780	1125	342	34.4
57	.357 MAG	158	JHP	SMITH+WESSON	4.00	1050	1116	340	14.4
84	.357 MAG	158	JHP	SMITH+WESSON	2.00	1050	982	299	11.1
33	.357 MAG	158	JSP(HI-VEL)	FEDERAL	4.00	1550	1255	382	21.1
41	.357 MAG	158	JSP(HI-VEL)	FEDERAL	2.00	1550	1195	364	18.7
47	.357 MAG	158	JSP	SMITH+WESSON	4.00	1500	1168	356	17.2
48	.357 MAG	158	JSP	SMITH+WESSON	2.00	1500	1091	332	17.0
23	.357 MAG	158	JSP	SPEER	4.00	1625	1156	352	28.0
46	.357 MAG	158	JSP	SPEER	2.00	1625	1030	313	17.5
50	.357 MAG	158	LRN(LUBALOY)	WESTERN SUP-X	4.00	1410	1230	374	16.6
58	.357 MAG	158	LRN(LUBALOY)	WESTERN SUP-X	2.00	1410	1169	356	14.4
91	.357 MAG	158	SWC	REMINGTON	4.00	1410	1088	331	10.2
97	.357 MAG	158	SWC	REMINGTON	2.00	1410	958	291	9.3
141	.38 SPEC	64	SHORT STOP	MBA	4.00	####	738	224	0.9
142	.38 SPEC	64	SHORT STOP	MBA	2.00	####	671	204	0.4
68	.38 SPEC	90	JSP(HEMI)	SMITH+WESSON	4.00	1350	1158	352	12.4
107	.38 SPEC	90	JSP(HEMI)	SMITH+WESSON	2.00	1350	1053	320	7.2
83	.38 SPEC	90	JSP	SMITH+WESSON	4.00	1350	1118	340	11.8
112	.38 SPEC	90	JSP	SMITH+WESSON	2.00	1350	975	297	6.5

BULLET ID NO.	CALIBER	WEIGHT (grains)	BULLET TYPE	MANUFACTURER	BARREL LENGTH (in)	VELOCITY NOMINAL* (fps)	VELOCITY MEASURED (fps)	VELOCITY MEASURED (mps)	RI INDEX
38	.38 SPEC	90	MP	KTW	4.00	1030	922	281	19.6
52	.38 SPEC	90	MP	KTW	2.00	1030	734	223	15.6
24	.38 SPEC	95	JHP(+P)	REMINGTON	4.00	985	1187	361	28.0
60	.38 SPEC	95	JHP(+P)	REMINGTON	2.00	985	1019	310	14.0
15	.38 SPEC	96	SAFETY SLUG	DEADEYE ASSOC	4.00	1800	1585	483	41.8
18	.38 SPEC	96	SAFETY SLUG	DEADEYE ASSOC	2.00	1800	1496	455	37.5
69	.38 SPEC	110	JHP	SMITH+WESSON	4.00	1380	1014	309	12.4
109	.38 SPEC	110	JHP	SMITH+WESSON	2.00	1380	888	270	6.8
89	.38 SPEC	110	JHP	SPEER	4.00	1245	857	261	10.5
106	.38 SPEC	110	JHP	SPEER	2.00	1245	789	240	7.7
27	.38 SPEC	110	JHP	SUPER VEL	4.00	1370	1159	353	25.1
28	.38 SPEC	110	JHP	SUPER VEL	2.00	1370	1148	349	24.8
40	.38 SPEC	110	JHP(LOT-Q4070)	WINCH-WESTERN	4.00	####	1106	337	19.3
61	.38 SPEC	110	JHP(LOT-Q4070)	WINCH-WESTERN	2.00	####	956	291	14.0
39	.38 SPEC	110	JSP	SUPER VEL	4.00	1370	1202	366	19.4
62	.38 SPEC	110	JSP	SUPER VEL	2.00	1370	1076	327	14.0
115	.38 SPEC	125	JHP	SMITH+WESSON	4.00	1350	900	274	5.9
136	.38 SPEC	125	JHP	SMITH+WESSON	2.00	1350	716	218	3.0
100	.38 SPEC	125	JHP	SMITH+WESSON	4.00	1350	1002	305	8.9
96	.38 SPEC	125	JHP	SMITH+WESSON	2.00	1350	899	274	9.6
37	.38 SPEC	125	JHP(+P)	SPEER	4.00	1425	1006	306	19.9
53	.38 SPEC	125	JHP(+P)	SPEER	2.00	1425	931	283	15.5
31	.38 SPEC	125	JSP(+P)	SPEER	4.00	1425	1047	319	22.5
66	.38 SPEC	125	JSP(+P)	SPEER	2.00	1425	983	299	13.2
56	.38 SPEC	125	JSP	SMITH+WESSON	4.00	1350	1064	324	14.5
117	.38 SPEC	125	JSP	SMITH+WESSON	2.00	1350	896	273	5.6
51	.38 SPEC	125	JSP	3-D	4.00	1085	1091	332	16.5

BULLET ID NO.	CALIBER	WEIGHT (grains)	BULLET TYPE	MANUFACTURER	BARREL LENGTH (in)	VELOCITY NOMINAL* (fps)	VELOCITY MEASURED (fps)	VELOCITY MEASURED (mps)	RI INDEX
92	.38 SPEC	125	JSP	3-D	2.00	1085	957	291	10.1
26	.38 SPEC	125	JHP	REMINGTON	4.00	1160	1108	337	25.5
108	.38 SPEC	125	JHP	REMINGTON	2.00	1160	911	277	7.0
42	.38 SPEC	140	JHP(+P)	SPEER	4.00	1200	978	298	18.6
43	.38 SPEC	140	JHP(+P)	SPEER	2.00	1200	897	273	18.5
71	.38 SPEC	148	WC	BROWNING	4.00	770	731	222	12.3
81	.38 SPEC	148	WC	BROWNING	2.00	770	618	188	11.9
70	.38 SPEC	148	WC	REMINGTON	4.00	770	741	225	12.4
74	.38 SPEC	148	WC	REMINGTON	2.00	770	700	213	12.2
72	.38 SPEC	148	WC	FEDERAL	4.00	770	737	224	12.3
75	.38 SPEC	148	WC	FEDERAL	2.00	770	674	205	12.1
73	.38 SPEC	148	WC	SMITH+WESSON	4.00	800	726	221	12.3
76	.38 SPEC	148	WC	SMITH+WESSON	2.00	800	662	201	12.1
77	.38 SPEC	148	WC	SPEER	4.00	825	679	206	12.1
80	.38 SPEC	148	WC	SPEER	2.00	825	652	198	12.0
78	.38 SPEC	148	WC(CLEAN CUTTING)	WESTERN	4.00	770	696	212	12.1
82	.38 SPEC	148	WC(CLEAN CUTTING)	WESTERN	2.00	770	618	188	11.9
67	.38 SPEC	158	JHP	SMITH+WESSON	4.00	1050	1047	319	13.0
88	.38 SPEC	158	JHP	SMITH+WESSON	2.00	1050	950	289	10.6
137	.38 SPEC	158	JSP	SMITH+WESSON	4.00	1050	828	252	2.9
140	.38 SPEC	158	JSP	SMITH+WESSON	2.00	1050	730	222	2.0
105	.38 SPEC	158	LRN	WINCHESTER	4.00	855	919	280	8.0
119	.38 SPEC	158	LRN	WINCHESTER	2.00	855	780	237	4.6
99	.38 SPEC	158	LRN(+P)	FEDERAL	4.00	1090	999	304	9.0
103	.38 SPEC	158	LRN(+P)	FEDERAL	2.00	1090	947	288	8.2
118	.38 SPEC	158	LRN	FEDERAL	4.00	855	795	242	5.0
127	.38 SPEC	158	LRN	FEDERAL	2.00	855	632	192	4.2

BULLET ID NO.	CALIBER	WEIGHT (grains)	BULLET TYPE	MANUFACTURER	BARREL LENGTH (in)	VELOCITY NOMINAL* (fps)	VELOCITY MEASURED (fps)	VELOCITY MEASURED (mps)	RI INDEX
120	.38 SPEC	158	LRN	REMINGTON	4.00	855	749	228	4.5
124	.38 SPEC	158	LRN	REMINGTON	2.00	855	694	211	4.4
121	.38 SPEC	158	LRN	SPEER	4.00	975	749	228	4.5
125	.38 SPEC	158	LRN	SPEER	2.00	975	635	193	4.4
126	.38 SPEC	158	LRN	SMITH+WESSON	4.00	910	708	215	4.4
135	.38 SPEC	158	LRN	SMITH+WESSON	2.00	910	626	190	3.5
44	.38 SPEC	158	LHP	WINCH-WESTERN	4.00	855	915	278	18.4
85	.38 SPEC	158	LHP	WINCH-WESTERN	2.00	855	805	245	11.0
86	.38 SPEC	158	SWC	FEDERAL	4.00	855	823	250	10.9
101	.38 SPEC	158	SWC	FEDERAL	2.00	855	796	242	8.5
131	.38 SPEC	158	SWC	SMITH+WESSON	4.00	1060	875	266	4.0
138	.38 SPEC	158	SWC	SMITH+WESSON	2.00	1060	678	206	2.5
87	.38 SPEC	158	SWC	SMITH+WESSON	4.00	850	1006	306	10.8
132	.38 SPEC	158	SWC	SMITH+WESSON	2.00	850	870	265	4.0
102	.38 SPEC	158	SWC	SPEER	4.00	975	803	244	8.5
116	.38 SPEC	158	SWC	SPEER	2.00	975	640	195	5.7
59	.38 SPEC	158	SWC	WINCHESTER	4.00	855	924	281	14.3
104	.38 SPEC	158	SWC	WINCHESTER	2.00	855	779	237	8.2
122	.38 SPEC	200	LRN	REMINGTON	4.00	730	647	197	4.5
133	.38 SPEC	200	LRN	REMINGTON	2.00	730	593	180	4.0
123	.38 SPEC	200	LRN	SPEER	4.00	850	710	216	4.5
129	.38 SPEC	200	LRN	SPEER	2.00	850	598	182	4.1
128	.38 SPEC	200	LRN(LUBALOY)	WESTERN SUP-X	4.00	730	626	190	4.2
130	.38 SPEC	200	LRN(LUBALOY)	WESTERN SUP-X	2.00	730	592	180	4.1
134	380 AUTO	95	FJ	WESTERN SUP-X	3.86	955	948	288	4.0

BULLET ID NO.	CALIBER	WEIGHT (grains)	BULLET TYPE	MANUFACTURER	BARREL LENGTH (in)	VELOCITY NOMINAL* (fps)	VELOCITY MEASURED (fps)	VELOCITY MEASURED (mps)	RI INDEX
3	.41 MAG	210	JSP	REMINGTON	4.00	1500	1260	384	51.9
65	.41 MAG	210	SWC	REMINGTON	4.00	1050	944	287	13.7
16	.44 MAG	180	JSP	SUPER VEL	4.00	1995	1495	455	41.6
1	.44 MAG	200	JHP	SPEER	4.00	1675	1277	389	54.9
8	.44 MAG	240	JHP	BROWNING	4.00	1330	1257	383	47.9
9	.44 MAG	240	JHP	REMINGTON	4.00	1470	1229	374	46.7
11	.44 MAG	240	JSP	SPEER	4.00	1650	1203	366	45.7
6	.44 MAG	240	SWC	BROWNING	4.00	1470	1311	399	49.8
7	.44 MAG	240	SWC	REMINGTON	4.00	1470	1286	391	48.9
5	.44 MAG	240	SWC	WINCH-WESTERN	4.00	1470	1330	405	50.0
114	.44 SPEC	246	LRN	REMINGTON	2.00	755	640	195	6.3
34	.45 AUTO	185	JHP	REMINGTON	5.00	950	895	272	21.1
55	.45 AUTO	185	WC(TARGETMASTER)	REMINGTON	5.00	775	821	250	14.7
95	.45 AUTO	185	WC	FEDERAL	5.00	775	751	228	9.7
110	.45 AUTO	230	FJ	REMINGTON	5.00	855	839	255	6.7
113	.45 AUTO	230	FJ	WINCH-WESTERN	5.00	850	740	225	6.5
111	.45 LC	225	LRN	WINCH-WESTERN	7.50	860	821	250	6.6

* = Advertized Velocity

Chapter 17

DEMONSTRATION OF TERMINAL BALLISTICS

IN my book co-authored with Charles Skillen,* I mention that it is wise in many instances to present a new training or ballistic concept to the press and to the city fathers so that they can grasp what the department is attempting to accomplish. Unfortunately, far too many departments fail to simplify the demonstration, thus accomplishing far less than desired. Most people are unable to understand the complexities of ballistics. They should not be required to attempt to understand them.

Several years ago, I put on a very simple demonstration before a group of top law enforcement officials and high civilian officials. The situation bothered me for some time prior to the demonstration. I wanted to make the demonstration simple, factual, and if possible readily understandable. I wanted to show how much safer and how much more effective high-velocity, hollow-point, jacketed bullets were than the customary conventional hard lead, round-nose, low-velocity bullets. I would also include the metal-point bullets and the armor-piercing bullets.

I knew that I would use Duxseal because it was a constant. On the other hand, after each shot, the Duxseal would have to be kneaded back into shape again. Because of the number of shots that would be fired, such a procedure would draw out the demonstration until it became boring. Something had to be done to quicken the entire operation.

We finally decided after a great deal of discussion to fill a dozen paper-wrapped one-quart oil cans with Duxseal. We also bought a large package of balloons. Next we built a small, lightweight, easily portable holder for the quart oil cans that

---

*Skillen, C. R. and Williams, M.: *American Police Handgun Training*. Springfield, Thomas, 1977.

would permit the cans to be set into place and held in place despite the bullets that would slam into the cans. We had to leave an opening behind the oil cans so that the balloons could be placed there.

Figure 17-1. Box with paper can of Duxseal and balloon in place. Bullet impact drives the can forward against the top cross bar and the raised plate at the bottom. The balloon is held in place by means of friction (between the top cross bar and bottom base plate). A bullet trap is placed behind the balloon.

In operation, the frame was set on a table top. Into the front of the frame was placed one of the quart oil cans filled with Duxseal. Behind was a balloon blown up as far as we dared. The larger the balloon, I might add, the more spectacular the results. The balloon could be filled with water or even a colored liquid, anything that would provide a visual destruction of the balloon. I was attempting the precise opposite of what is normally expected of a bullet testing apparatus. I wanted to dramatically illustrate the penetration capabilities of the slow-moving, hard-lead, round-nose handgun bullet. However, tied into this was a concurrent demonstration of the safety of the high-velocity, jacketed, expanding bullets.

During the demonstration I would first permit an examination of the cartridge to be fired along with a brief series of comments explaining the cartridge, its velocity, and bullet construction. Using a Smith and Wesson four-inch barrel .357 Magnum revolver, I would then load that specific cartridge into the revolver chamber, step back to about fifteen feet and fire into the center of the Duxseal. The bullet would remain within the Duxseal. I would pass the can around and let the various people examine it. I would then take the same can, replace it in front of the balloon turning it so that the bullet channel from the first shot was not facing me and fire another shot. This time, again after passing the cartridge around to be inspected, I would fire the hard-lead, round-nose bullet into the Duxseal. Such a bullet will easily pass through the oil can and demolish the balloon. Again the can would be passed around. Someone would inevitably — and I wanted this to happen — complain that "sure, the Duxseal had already been opened up and no wonder the bullet went all the way through."

I would then pass around a new Duxseal-filled can along with another lead bullet loading then set them up with a new balloon and fire again. When the balloon again broke, most of the watchers were convinced. Then in order to nail down the situation, I would run through a series of shots, quite often using an officer's handgun and his duty ammunition in both .38 Special and .357 Magnum, thus permitting the viewers to make up their own minds.

This is a simple demonstration but one that has proven to be extremely effective particularly when done in front of a large group of people.